Naturalist's Color Guide Supplement

NATURALIST'S
COLOR GUIDE
SUPPLEMENT

FRANK B. SMITHE

The American Museum of Natural History
New York

The American Museum of Natural History
New York, New York 10024

Contents

Foreword

Sight is probably the most important of the five senses normal to mankind; it surely is for the biologist. And of the elements of sight, the ability to distinguish colors is crucial. Animals and plants come in a bewildering variety of colors and color patterns. These are often adaptations for various purposes—courtship, concealment, warning signals, species recognition marks, and others.

To study such biological phenomena and to describe and compare species and populations accurately, the scientist needs a convenient color reference, tailored to his research requirements. Complicated devices now exist for measuring the physical properties of light, and hence of color. Such equipment is used by biologists in some situations, but it is too cumbersome or otherwise unsuitable for many purposes. The plumage of a bird or the fur of an animal is rarely a simple, uniform patch of color. Hairs and feathers vary in color from base to tip; they form an uneven surface with highlights and shadows. Individual variation may be so great that rendering too precise a measurement of one specimen could be misleading. What is needed is a color reference that permits direct visual comparisons to a degree consistent with the inherent properties and variability of the animals or plants being studied.

The most ambitious effort to provide such a reference for naturalists and systematists was that of Robert Ridgway (1850–1929), longtime curator of birds at the Smithsonian Institution. His *Color Standards and Color Nomenclature* (1912) contained swatches of 1115 shades and colors, many of them named by himself. This was, quite simply, too many colors. Ridgway himself, writing eight volumes of *The Birds of North and Middle America,* never used some of the colors he had provided for his own work; in an astonishing number of other cases he did not mention a specific color without some modifier—thus, light sepia, dark sepia, deep sepia, etc. With over one thousand colors he still found it necessary to modify. In any case, Ridgway's color guide is now so rare, and the evidence of deterioration in some of the colors so convincing, that almost no one attempts to use it, although many of the names for colors that he introduced continue to be used.

A number of other color guides have been published—some very elaborate—such as those of Munsell, Villalobos, Ostwald, etc.

However, these were compiled for other purposes, often for trade or industry; the names used for the colors are often bizarre or contradictory; the works themselves are often rare, expensive, unreliable, or all three.

When my friend Frank B. Smithe came to me and announced his plan to work out and publish a color guide for the use of naturalists, I was aware of at least some of the potential difficulties and sought to discourage him. I should have recalled the unflagging diligence, patience, and judgment he brought to his earlier work in ornithology and archeology and bade him Godspeed. He has pursued these avocations after retiring from business, with training in science and engineering, and his volume *The Birds of Tikal* stands as evidence of his unusual abilities.

The present contribution is offered in two parts. First is the NATURALIST'S COLOR GUIDE itself, in which swatches of eighty-six colors are provided, along with an ingenious mask that permits each color to be singled out without the disturbing influence of an adjacent swatch. Each color is accompanied by a name, a number, and a numerical notation of its measurement, both before and after printing. The NATURALIST'S COLOR GUIDE thus avoids the questions haunting earlier color keys: "How much has this color changed over the years?" "How does this copy of the book compare with others?" To ensure quality further, the deposition of color lacquers has been done to exacting specifications, under the most rigorous of controls.

The second and longer part, the NATURALIST'S COLOR GUIDE SUPPLEMENT, will be valuable to those having a special interest in the subject—or certain aspects of it. Smithe has retained as much of Ridgway's terminology as possible, and he explains at length how the selections and simplifications were made. It is not always possible to be sure just what a Ridgway color is by referring to his color key, copies of which are now over sixty years old (there was also a bootlegged inferior reprint). Therefore, Smithe went through the eight big volumes of *The Birds of North and Middle America* line by line, finding where and for what birds Ridgway used each named color. As noted above, this was often easier said than done—Ridgway created some colors he never used, and he was very cautious about using others.

The next step was to examine carefully series of the species of birds in question under standard conditions of lighting at The American Museum of Natural History. Only then was the final selection of colors made.

Frank Smithe's color guide thus represents an immense amount of painstaking labor. Since he is too modest to say so himself, I should add that, like several of his other projects, the NATURALIST'S COLOR GUIDE represents a very substantial financial investment, the proceeds of which, as always, go to one or more of the cultural institutions fortunate enough to number Frank Smithe among their benefactors.

DEAN AMADON
Lamont Curator of Birds
The American Museum of Natural History

Preface

While writing descriptions of birds for *The Birds of Tikal,* I found it essential to examine similar descriptions made by others in a broad range of books and papers. In this study I discovered substantial confusion among ornithologists when they named the colors of plumage and other parts of birds. It was very apparent that there was no accurate guide to colors available, and one was obviously needed.

A considerable research effort was required to determine how best to fill that need—what colors must be included, how few rather than how many would be adequate, and so forth. The research indicated that Robert Ridgway did not eliminate the confusion, even though his color guides include more than a thousand colors and color names. It showed that inclusion of all the color names used by ornithologists would only compound the problem. For these reasons (and others), it seemed best to limit severely the quantity of color names to those deemed most useful for naturalists.

After more than four years of concentrated work, the NATURALIST'S COLOR GUIDE resulted. It will, at the very least, prescribe a specific color name or color number for the colors it depicts. For these colors, confusion should no longer exist.

INTRODUCTION

Introduction

The objective of the NATURALIST'S COLOR GUIDE is to give ornithologists and other naturalists an accurate and handily available tool to use in describing the colors of birds and other animals. Part I presents a series of eighty-six colors, arranged along the outside edge of the page, whereby a bird's plumage (or any other object) can be viewed and compared with the color swatch. Each color has a name, a number, and a numerical notation. Part II, the SUPPLEMENT, gives the details of these factors. This introduction begins with an explanation of their derivation from the works of Robert Ridgway.

Robert Ridgway

Nearly one hundred years ago Robert Ridgway, while curator of birds at the United States National Museum, started to accumulate material for a series of books entitled *The Birds of North and Middle America.* The first volume was published in 1901 as Smithsonian Institution Bulletin No. 50, Part I. Seven more volumes followed, Part VIII being published in 1919. Ridgway died in 1929, and his work was continued by Herbert Friedmann, who issued Part IX in 1941, Part X in 1946, and Part XI in 1950. Two additional volumes were projected to include the species from Tinamiformes (tinamous) through Anseriformes (ducks, geese, swans), but the project has been dropped.

The eleven volumes of Bulletin 50, as it is hereafter called, are a remarkable work. They describe in great detail the external characteristics and plumage of every species then known in North and Middle America (with the exceptions noted above). The NATURALIST'S COLOR GUIDE SUPPLEMENT includes the scientific and vernacular names of some of the birds that, according to Ridgway, display certain colors conspicuously in their plumage. These citations are chosen from Bulletin 50, but the Bulletin presents two problems. First, the sequence of species is in reverse order from that of present guide usage, and second, the nomenclature is frequently outmoded. Therefore, a comparative list of Ridgway's nomenclature versus modern nomenclature, in reverse order from Bulletin 50, is shown in Appendix B of the SUPPLEMENT.

Ridgway wrote two other books, in which he defines the colors of the plumage of the birds so meticulously described in Bulletin 50. *Nomenclature of Colors . . . for Ornithologists* was published in 1886. Hereafter referred to as *Nomenclature,* this color reference book has

3

The Birds of North and Middle America

Vol.	Author	Year Published	Mention of 1886 *Nomenclature*	Mention of 1912 *Color Standards*
I	Ridgway	1901	pp. xvii, 289, 291	—
II	Ridgway	1902	pp. 315, 408, 703	—
III	Ridgway	1904	pp. 6, 397, 716, 748, 749	—
IV	Ridgway	1907	pp. 52, 179, 748	—
V	Ridgway	1911	pp. 7, 787	—
VI	Ridgway	1914	p. 119	pp. 135, 783
VII	Ridgway	1916	p. 122	p. 424
VIII	Ridgway	1919	—	—
IX	Friedmann	1941	—	—
X	Friedmann	1946	—	—
XI	Friedmann	1950	—	—
XII, XIII	—	—	—	—

ten color plates and shows 186 samples of named colors. Ridgway frequently cites *Nomenclature* in the first seven parts of Bulletin 50 (see above table).

In 1912 Ridgway published a much more ambitious compendium of colors, which has been a standard guide for many ornithologists ever since: *Color Standards and Color Nomenclature,* hereafter referred to as *Color Standards.* It has fifty-three color plates and shows 1115 swatches of named colors. *Color Standards* is first cited in Part VI of Bulletin 50 (1914) and then in Part VII (1916) (see table). Evidently, Ridgway relied only on *Color Standards* from the time of its publication, but he seems to have used the unpublished work before 1912. He wrote a letter to *Ibis* (Vol. III, p. 714), dated September 9, 1909, which says in part:

> Probably some of the readers of "The Ibis" are aware that I have been engaged, from time to time as opportunity allowed, on a new and greatly improved edition of my Nomenclature of Colors (1886). I am happy to be able to announce that, after twenty years of necessarily intermittent labour, this most difficult and tedious task has at last been accomplished.

It is probable that Ridgway had been using his color material even before this letter; earlier plumage descriptions support this inference. Most of the colors mentioned in Bulletin 50 agree with the swatches in *Color Standards.*

Color Standards agrees in the main with *Nomenclature,* with two broad exceptions: many new colors are added, and thirty-four former colors are omitted or replaced. The addition of new colors does not cause any confusion, but this cannot be said of the omissions. Many of

4

the omitted colors appear frequently in early descriptions, which results in some confusion. A few examples are worth noting:

1. Broccoli Brown is not shown in *Color Standards,* although it is often used in Bulletin 50. It may be close to Drab and Hair Brown, which are shown in both *Nomenclature* and *Color Standards.*
2. Canary Yellow seems to have been dropped in favor of Citron Yellow.
3. Chrome Yellow and Saffron Yellow are dropped, perhaps in favor of Deep Chrome or Cadmium Yellow.
4. Cobalt Blue is dropped, perhaps in favor of Ultramarine Blue.
5. Crimson is dropped, probably in favor of Carmine.
6. Ochraceous is dropped, probably replaced by Ochraceous Orange.
7. Poppy Red is dropped, after frequent early use, and it is not possible to identify an equivalent.
8. Tawny Ochraceous becomes Ochraceous Tawny.
9. Vermilion is dropped, probably in favor of Scarlet.
10. Violet is dropped, but Amethyst Violet is retained.

On the other hand, an example of Ridgway's consistency is worth quoting. The very first species he describes is the Evening Grosbeak, *Hesperiphona v. vespertina,* in Bulletin 50, Vol. I, p. 39:

Adult male. Forehead (more or less broadly) and superciliary region yellow; rest of pileum black; rest of neck and upper back plain olive, lighter and more yellowish on throat, changing gradually to clear lemon yellow on scapulars and rump and to lighter (more citron) yellow on posterior underparts, the longer under tail-coverts sometimes partly white; upper tail-coverts and secondaries (tertials) which are white or pale grayish, the former sometimes edged with yellow; bill light olive-yellowish or pale yellowish green.

A little further on (p. 47) he describes the adult male Cardinal as "general color dull red (varying from dull brownish scarlet or almost orange-chrome in summer to a hue more or less approaching dragon's blood red in winter)."

Such meticulous descriptions, published in 1901, with only slight changes agree with the color names found in *Color Standards.* Olive-Yellowish became Olive-Yellow in 1921; Yellowish Green became Yellow-Green; Citron became Citrine; Orange-Chrome lost the hyphen; Dragon's Blood Red gained a hyphen.

The passage of time has created another serious problem. In 1909 Ridgway wrote that the colors were to be reproduced with "as great a degree of permanency as is possible with pigments now known." Regrettably, in the course of time many of the color swatches in *Color Standards* have lost their original hue, their lightness, or their saturation. This fact is evident when different copies of the original book are examined and compared with each other. Some copies display the deviations more than others. My own copy shows obvious deterioration, even without comparison to other copies. Some neighboring color swatches are not even separable into distinctly different colors. No doubt other copies have been better preserved, but the pigments used around 1912 were not as permanent as was then hoped, and many of the color swatches in even the best preserved copies have now deteriorated. Also, it is not possible to tell beyond question which swatches have *not* deteriorated.

In 1948, John T. Zimmer said in *Science* that in his opinion this deterioration had not in fact occurred. He thought that the deviations were due to reissues of *Color Standards* in 1937 and 1940, which lacked identification that they were reprints. His entire argument is as follows; perhaps others will agree with him:

The account in Science, Vol. 107 (June 11, 1948, pp. 626–628) by Illman and Hamly on the unreliability of Ridgway's classic volume was very interesting. The authors seem not to have realized, however, that there is more than one edition of Ridgway's work. Some of the data they tabulate appear to concern copies of different editions and hence give somewhat false criteria by which to judge the imperfections of the book. The particular copy that they call the "good copy" is almost certainly one of the reprints, and perhaps others of the series examined are of the same sort.

Presumably because of the continuing demand for Ridgway's work, after the edition was exhausted, the printers of that volume undertook to reissue the work about 1937, using an undetermined number of leftover sheets and preparing new ones to fill the gaps—without the benefit of Ridgway's personal attention, since he had died some years previously. Still later, about 1940, an entirely new set of plates was projected, but whether they were issued or not I am unable to say. Unfortunately, no indication was given on the title page or in the letterpress that the new books were different from the originals, although there are minor distinctions that are apparent on comparison with an original, other than those found in the colors themselves. The colors in many cases are far from accurate counterparts of those of Ridgway's own preparation.

Of two original copies immediately available, one has been very little used, while the other has seen continuous service for the last 18 years. Although both show some spotting and discoloration—much more evident in the heavily used copy—the unaffected portions are identical in both or so nearly alike that a colored object matched in one set would find the same place in the other. On the other hand, a relatively new copy of the reissued work is decidedly different from the others. The differences, I am sure, are not due to deterioration of the older examples but to faulty preparation of the reissues.

I have no wish to criticise the *Munsell Book of Color* of which Illman and Hamly speak so highly. It is indeed a fine work, and one may hope that the pigments used in its preparation will prove to be more permanent than some of those available to Ridgway in 1912. I maintain, however, that Ridgway's *Color Standards and Color Nomenclature* has not deteriorated so much and is not so useless as we are asked to believe. The problem is one of keeping references restricted to the original work, which I admit is difficult, since the printers have given no unequivocal clue to the reprints as they should have done. Whether the reprints are uniform among themselves also remains to be determined by someone with access to a number of copies. At any rate, they should not enter into a judgment of the original work.

Even if Zimmer could say that copies of the 1912 *Color Standards* were uniform when compared in 1948, that does not mean that some of the colors had not changed materially in the intervening thirty-six years. Today, after another quarter of a century has passed, certainly all copies have more or less deteriorated since 1912. For this reason, and because of the increasing rarity of the work—not to mention the excessive numbers of colors it included—*Color Standards* is very rarely used. But many of the colors and color names Ridgway introduced are still in use, and hence his work served as the logical point of departure for this work.

Another problem of *Color Standards* is that it lacks precise descriptions of how to reproduce its colors. While Ridgway did have a system, it was based on the use of pigmented color disks, thirty-six having a specific color and the others black, white, and shades of gray. The disks were spun around in combinations that blended to produce the color he desired. The procedure is fully described in *Color Standards,* but the task of duplicating the combinations today would probably be impossible, especially in view of the fact that we would not know exactly what colors we aimed to produce, not knowing for certain what colors Ridgway intended in *Color Standards.*

Another important fact must be considered when attempting to match a color description from Bulletin 50 with *Color Standards* and with a specimen. Different individuals will find different matches; that is, one person will visualize a plumage color and match it with a color swatch and arrive at a different result from that of another person. To some extent, this difference may be a matter of the age of the viewer; a young person's choice of a color match will differ from that of a substantially older person. To some extent, it is an inherent difference; some people simply see colors differently from others. And to some extent, it may be due to the obvious physical difference between a plumage surface and a painted swatch surface.

Color Standards shows 1115 named colors. This enormous number of swatches is obviously not suitable for a handy color guide. Ridgway himself actually uses only a small portion of his colors, and there is often a dearth of useful citations in Bulletin 50. With so many colors to choose from, Ridgway more often than not specifies a plumage in terms of two or more colors or in terms of a range of colors. While in some instances he may be indicating that the species described varies individually or seasonally within the color range, he is often indicating that the color is neither one nor the other, but some color between the two. When he does use a color, he often modifies it, saying it is "dark," "light," "near," or whatever.

Ridgway's work is of outstanding importance to ornithologists, but because of the many problems inherent in his color guides, several other guides and color systems have been developed over the years. Some of them were very helpful in compiling the NATURALIST'S COLOR GUIDE and are discussed below.

Ralph S. Palmer

In 1962 Ralph S. Palmer published *Handbook of North American Birds,* which includes a color chart of only twenty-one named colors. It also includes a hexagon of chromatic colors. The beautiful plates of Palmer's color chart are about the only effort since Ridgway's to publish a color guide for ornithologists, and Palmer's colors are referred to throughout this guide, which uses nearly all of Palmer's color names. In some instances the choice of color swatch for the same color name in the NATURALIST'S COLOR GUIDE does not agree with Palmer, but there are instances of complete agreement. Any differences are a matter of choice rather than criticism. Palmer's chart has been extremely useful in my work, and I gratefully acknowledge his cooperative spirit in our discussions.

David H. Hamly

David H. Hamly made a series of measurements of the color chips in Ridgway's *Color Standards.* In 1949 he published an alphabetical list of the color names and notations of *Color Standards* and gave the corresponding Munsell notations from his measurements. This correlation is a sort of Rosetta Stone to Ridgway's colors in terms of Munsell notations, and Hamly's notations are quoted throughout the NATURALIST'S COLOR GUIDE. The accuracy of Hamly's measurements is largely dependent, of course, on the condition of the copy of *Color Standards* he used; some of his notations are open to question.

Villalobos Color System

In 1947 the *Altas de los Colores,* by C. Villalobos-Dominguez and J. Villalobos, was published in Buenos Aires. The Villalobos color system was used by Palmer to produce the color chart tipped into his 1962 work. The atlas shows 7279 printed colors and includes an appendix of equivalents for the colors in Ridgway's *Color Standards.*

8

Villalobos designates an individual color in written symbols: a letter or letters denotes each of thirty-eight basic hues and its relative position in the spectrum; a number denotes the color's position in a scale of nineteen divisions of lightness between black and white; another number with a degree mark denotes the percentage of intensity (called chromaticity) in twelve steps from weakest to strongest. Rose color, for example, reads MR-12-11°.

While an excellent system on which to base a color guide, the Villalobos symbols primarily describe the position of a color on a color chart rather than a specific measurement of a color. Such measurements are found by using the Munsell system.

Munsell Color System

The Munsell color system is scientifically designed to define a color accurately. It divides a color into three aspects: the spectral color called hue, the degree of lightness or darkness called value, and the intensity or saturation called chroma. Each aspect is given a symbol, and the combination results in a Munsell notation identifying the color. The Munsell system enables one to make a mental picture of the color, an advantage not found in most color systems. The NATURALIST'S COLOR GUIDE relies heavily on the Munsell system, which is described briefly here and is fully described in literature available from the Munsell Color Co. in Baltimore.

The hue notation is symbolized by capital letters indicating the relative position of the hue in a circle of the ten major colors of the visible spectrum. The symbols are R for red, YR for yellow-red, Y for yellow, GY for green-yellow, G for green, B for blue, PB for purple-blue, and RP for red-purple. A number preceding the letters indicates the position of the color within the range of its hue, numbered from 1 to 10 for each of the major hues. For example, in the range of red hues, a red almost red-purple is 1R, middle red is 5R, and red almost yellow-red is 10R or 1YR.

The value of a color is the degree of lightness or darkness related to a scale of gray, from black to white. The symbol 0/ denotes absolute blackness, 10/ denotes absolute whiteness, and 5/ denotes a middle gray, or colors that appear halfway between black and white.

The chroma notation specifies the intensity or saturation of the hue. A color with a weak chromaticity will have a notation of /1, a moderate chroma will be /6 or /8, and a vivid, saturated hue may have a chroma notation of /14, /16, or even higher.

The combined Munsell notation is then written in terms of *hue value/chroma*. Spectrum Red, for example, has a notation of 5R 4/ 15. It can be visualized as a color with a middle red hue (5R), a value

(4/) about midway between black and white, and a chroma (/15) of very high intensity or saturation.

Finer divisions of these aspects are expressed in decimals (for instance, a specimen of the Great Crested Flycatcher was measured carefully and given a notation of 4.8Y 3.8/1.8), but such a fine degree of accuracy is not essential to a color guide such as this.

Ostwald Color System

The Center for Advanced Research and Design (a division of the Container Corporation of America) has published the *Color Harmony Manual,* based on the Ostwald system. It has nearly one thousand colors made of lacquers on plastic hexagonal chips. Each color chip has a glossy surface on one side and a matte surface on the other. Munsell notations for the colors in the third edition were determined by spectrophotometric measurement and privately published by Walter C. Granville in 1948. The *Color Harmony Manual* is supplemented by a dictionary with a color name for each color chip. While Ridgway's *Color Standards* is listed as a reference for the dictionary, most of the color names are geared for the mass market and seldom correspond to those used by ornithologists. Even when a Ridgway name is given, the color chosen to represent it does not agree with Ridgway's swatch for that color.

Methuen Handbook

The *Methuen Handbook of Color,* by A. Kornerup and J.H. Wanscher, is of special interest, partly because it lays great stress on color names and partly because it tabulates each of its 1260 color prints directly into equivalent Munsell notations. The handbook comprises thirty plates of colors with forty-two colors on each plate. Methuen notations identify each color by plate number, followed by a letter locating its horizontal position on the plate, and ending with a number denoting the vertical position on the plate. A tabulation correlating each color with its equivalent Munsell notation is also included.

Winsor & Newton Color Chart

Winsor & Newton published a color chart entitled *Artists Water Colours.* Although it is directed primarily to use by artists, its concentrated display of vivid colors could be of value to ornithologists. The chart has only sixty colors, which is one of its advantages. Each color swatch shows a range of tints from the most saturated or full color to the palest or most washed-out color. The swatches do not appear to have been printed in multiple-color lithography, because they can be examined very closely and from any angle without alteration of their quality. Each color has a name assigned to it, but, as might be expected, few colors or color names correspond with those of Ridgway in *Color Standards*. The colors were measured spectrophotometrically for the NATURALIST'S COLOR GUIDE, and the resulting Munsell notations are quoted in the text when their use seems appropriate.

U. S. Bureau of Standards

The U. S. National Bureau of Standards has published a dictionary of color standards (Kelly and Judd, 1955). Supplemental to it is a set of color charts of excellent quality but limited usefulness for ornithologists. Eighteen pages show 251 colors, and each color is tabulated with its equivalent Munsell notation.

Miscellaneous Guides

Many other color guides have been devised, usually for industrial or other special purposes—textiles, ceramics, cosmetics, paints, printing inks, etc. A study of many of the available guides and the literature on color emphasizes the types of problems which even a limited guide must try to solve, such as producing a guide accurately after its desired colors have been determined. One commercial company has developed a guide for printers. It comprises two sets of colors printed on looseleaf pages from which chips may be removed. One set is printed on coated paper and the other on uncoated paper to show the visual difference between the two. Each color has an identifying number and a formula that tells the printer just what mix of inks is required to produce that color. The fact that the same mix of inks looks very different on coated and uncoated paper stresses one of the problems facing a printer when he is required to produce accurate and authentic colors.

Color can be measured by a variety of colorimeters and spectrophotometers. The most accurate instrument is the recording spectrophotometer, which measures the spectral aspects of color in terms of wavelengths of light and in terms of the relative reflectance or energy recorded for each wavelength. It physically delivers a diagrammatic curve that pictures the results of the measurements, with wavelengths and reflectance percentages as coordinates.

When a computer is combined with a spectrophotometer, additional information is delivered. A series of numbers results, representing what are known as tristimulous values designated as

11

large X, Y, and Z. From these, two additional numbers are derived, called chromaticity coordinates, designated as small x and y.

Data from such instruments can be converted to an international color specification system—that of the Commission Internationale d'Eclairage (CIE). The CIE system is correlated with the Munsell system by means of a series of conversion tables and diagrams available from the Munsell Color Co. The conversion method is described in ASTM Standard Method D-1535 (American Society for Testing & Materials, Philadelphia). The ability to convert from CIE to Munsell and the reverse, coupled with spectrophotometric measurement, allows color data to be recorded and used to determine the effect of aging on colors.

Color Terminology

Color is defined, for the purpose of this guide, as the interpretation by the human brain of the light reflected to the eye from any object. This definition includes black and white as well as the full range of hues.

Hue is the basic characteristic of a color—that aspect locating it in the sequence of colors based on the visible spectrum of sunlight (plus nonspectral purple and magenta). As noted by Evans, there may be some ten million distinguishable hues. This guide accepts a division of the spectrum into ten named hues, with further numbered divisions into finer relative positions in the spectral sequence.

Value is the lightness of a color, ranging from black to white. Such terms as *dark, dusky, light,* and *pale* are used to modify *value.*

Chroma is the intensity or saturation of a color. *Vivid, saturated, intense, full, strong, rich, unsaturated, weak,* and *dirty* are used to modify chroma.

Deep, while retaining the sense of a strong chroma, also includes a feeling of darkness or decrease in value. A color called *very deep* is thought of as even darker than *deep.* The word may derive from the change in quality or tone when a thin layer of a colored liquid is compared with a thicker or physically deeper layer of the same liquid (see Maerz and Paul, p. 9). In a Munsell chart of comparative colors, the lightest swatch in a series of the same hue (for instance, a reddish color with a 5R notation) would have the lowest chroma and the lightest value, with a possible notation of 5R 9/1. The darkest swatch of the same series would have the same low chroma (/1) but the maximum of darkness and a possible notation of 5R 2/1. The ambiguity of terms such as *deep* is precluded by the use of the notation.

Tint is used by Ridgway to designate the paling effect on a hue by the addition of whiteness; he indicates progressive additions of white as "a medium, light, pale, or delicate (pallid) tint" of hue.

Shade is used by Ridgway to indicate the darkening effect on a hue by the addition of black—the opposite of tint. He says, "a medium, dark, or very dark (dusky) shade" of hue.

Tone (of a hue or color) combines the terms *tint* (in the direction of lightness) and *shade* (in the direction of darkness); *tone* is related to *value,* which is preferred, and use of *tone* should be avoided.

The **spectrum** is formed from sunlight. When limited to a discussion of color, it is the visible portion of the radiant energy reaching us from the sun. If sunlight is transmitted through a prism, its rays are refracted and dispersed. If the dispersed rays are then reflected to the eyes from a flat surface, the resulting series of colors is the spectrum. Rainbows, which exhibit the colors of the spectrum, are caused by the refraction and reflection of the sun's rays by raindrops in the atmosphere, becoming visible when seen from a favorable angle.

It is common practice to identify the colors of the spectrum by six specific color names: red, orange, yellow, green, blue, and violet. It is also customary to measure the rays of light in terms of wavelength (measured in millimicrons, or mμ, or nanometers) and frequency. The visible spectrum ranges from about 760 mμ (the long-wavelength, low-frequency, red end of the group) to about 380 mμ (the short-wavelength, high-frequency, blue or blue-violet end of the group). Violet is composed of a combination of short-wavelength blue and long-wavelength red and may not be visible as true violet in a spectrum. The theories of nonspectral colors at this end of the hue range are not considered here.

Some authorities avoid the use of spectral color names, reasoning that such names as red or green are meaningless. They prefer calling a color scarlet or geranium, or even a fanciful name such as Eugenia Red (Ridgway) or Alizerin Crimson (Winsor & Newton). Winsor & Newton do not use any of the spectral colors as color names. Palmer uses orange, yellow, and green as color names, but not red or blue. Ridgway uses orange and prefixes three others with the term *spectrum*: spectrum red, spectrum blue, and spectrum violet. Methuen uses all six spectral colors, prefixing each with the term *spectrum*. He also specifies the wavelength and Munsell notation of each color. The names of the colors of the spectrum are as useful as any names, provided the term *spectrum* is prefixed and a Munsell notation given. The suffix *-ish* should be used for all spectral colors when used in a generalized sense. Spectrum red, for instance, is then the name of a specific red, while a generalized red is called reddish.

Primary colors are those from which, when selectively mingled, all other colors can be produced. The accuracy of this definition, however, depends on whether one is mixing light (energy) or pigments (paints). For mixtures of light, such as reflectances from a whirling color wheel, the primary colors are red, green, and blue. For mixtures of pigments, the primaries are red, yellow, and blue. There are other complications, partly related to photography and/or projection of colors on a screen.

Warm colors include the range of hues from red through yellow.

Cool colors include the range of hues from green to violet, but some people restrict them to green-blue and blue hues.

Paired colors, in accord with usage in the English language, should follow a definite rule: the first named color modifies the second. For example, orange yellow denotes a yellow with an orange quality; olive green is a green of an olivaceous nature.

Determining a Color

The following discussion gives the details of the selection, naming, and measurement of a color for the NATURALIST'S COLOR GUIDE, using Olive as an example. Olive is included in the COLOR GUIDE partly because it is clearly and understandably used by Ridgway in Bulletin 50, partly because it is frequently used by ornithologists, and partly because it is closely related to other colors combining its name, such as olive brown, olive green, olive gray, etc.

But what is the correct color for Olive? If it is that of the olive fruit, is it a green olive, a ripe olive, a cocktail olive, a chemically treated olive, or what? Just what color do people mean when they describe a plumage as olive? Matching a color name, when it is finally chosen, to a color swatch is extremely difficult, and there are many different versions of Olive in the literature.

Hamly gives Ridgway's Olive, found on Plate XXX 21'm of *Color Standards,* a Munsell notation of 5Y 3.8/3.0. Another person's test on a different copy resulted in a 7.5Y 3.5/2.3 notation, practically an olive green. My own copy measured 2Y 3.6/2.0, a brownish olive hue. The range of hue of these three texts is indicative of the serious differences in various copies of *Color Standards,* now over sixty years old.

Olive swatches by other authors also show lack of agreement. One author's swatch has a 2.5Y 3/4 notation, similar to that of Ridgway's Brownish Olive (2.5Y 3.8/3.0 as measured by Hamly). Another author makes his own correlation for olive, giving it a notation of 10Y 3.7/3.1, but stating that olive is "a general name typified by the

14

sample indicated, which also represents olive green." Still another author gives up the ghost entirely, saying that "the sole distinction between the two terms today is that olive green is generally used in England, while olive is preferred in America." A swatch he calls olive green measured close to 5Y 4/3, which is similar to Ridgway's Olive.

It should be noted that, while there is a wide variance of hue among the various authors mentioned above—ranging from brownish olive through olive to greenish olive— there is close agreement on the value (from 3/ to 3.8/) and chroma (from /2 to /3.1, with one exception); all agree that Olive has a strongly grayish value and a rather low chroma.

In order to determine what Ridgway's Olive really is, many museum skins of birds described unequivocally as olive in Bulletin 50 were studied both visually and mechanically. Visual comparisons were made against swatches by Munsell, Villalobos, Ostwald, and Methuen. The Munsell Color Co. made (to special order) more than four hundred color swatches with extremely small color variations from brownish olive to greenish olive. There were many combinations of hue, value, and chroma, ranging in a narrow field from 0.2Y to 10Y in hue, 2.8/ to 5.2/ in value, and /0.4 to /4.5 in chroma.

All visual comparisons were made under uniform, carefully controlled light conditions. A special booth excluding external light was painted a neutral gray (Munsell's N 7 to N 8) to eliminate competing colors and to produce high visibility. Macbeth "daylight" lamps (Norlite Model NL-D75-440) with a color temperature of 7430° K were set at a distance from the work area to give an intensity of 150–200 footcandles on the objects measured. The color swatches and bird specimens were viewed at about ninety degrees from the line of sight by tilting them up about forty-five degrees to the lamps above them.

Spectrophotometric measurements were made under equally carefully controlled conditions at the laboratory of Henry Hemmendinger in Belvidere, New Jersey. The data from the measurements were very helpful as a crosscheck against visual measurements, but such measurements should be used with caution. When the plumage tested is clearly of a uniform color, of substantial area, and not bulged into the aperture of the instrument, the resulting measurements are acceptable. However, these conditions are not always present. In a plumage of mixed colors, especially when subdued by underlying brown or gray, the results can be questionable. The instrument records what it sees, so to speak, measuring the intermixing or background color as well as the vivid hues. Visual observation, on the other hand, tends to select these vivid hues, preferring those one wishes to describe, and the eye may even be physically unable to see a blending of hues when the specimen is in hand. At a distance, the eye tends to "choose" to see the more vivid hues—perhaps a psychological phenomenon.

It is also probable that the texture of the features reduces the value and chroma measured by the instrument. There was evidence that the very vividly colored plumages measured substantially darker than was recorded visually. Some of this loss of intensity may be caused by the angle at which the specimen is presented to the instrument. This is especially true when iridescence measurement is made; every slight change of angle produces a different result. Such measurements should hardly be attempted.

The birds used in the study of Olive were specimens of the Great Crested Flycatcher, *Myiarchus crinitus*. Ridgway (Vol. IV, p. 613 of Bulletin 50) describes them as "above plain olive, the pileum usually browner." The word *above* refers to an area including the nape, back, scapulars, and upper rump—a substantial area of uniform color. The word *plain* indicates that the area is uniform and not conspicuously streaked. Blake describes this area of the Great Crested Flycatcher as dark or dull olive; Bent says olive brown; Land says brownish olive; Smithe says olive green; Imhof says olive green; Chapman says grayish brown washed with olive green; Hausman says grayish olive brown. Only Ridgway says olive without reservation.

Birds that do not change their plumage seasonally will usually duplicate their plumage very closely every time they produce a fresh one; after a period of wear and tear the plumage may look somewhat different. Ornithologists at The American Museum of Natural History said that the late summer and early fall would be the best time to find the Great Crested Flycatcher in fresh plumage. Therefore, it was essential to measure specimens in spring or summer plumage separately from those in fresh autumn plumage.

Two specimens, nearly identical in color, were selected from many in the Museum's collection labeled with a spring date. Visual tests compared the specimens with the four hundred specially prepared Munsell swatches, giving the following results:

1. Specimen 104317, dated June 28, 1906, was tested thirty-five times and averaged 2.2Y 3.8/2.0.
2. Specimen 369252, dated May 1915, was tested twenty-four times and averaged 2.5Y 3.9/2.1.

The average of all fifty-nine tests was 2.32Y 3.82/2.06, which checks fairly well with Hamly's correlation of Ridgway's Brownish Olive swatch at 2.5Y 3.8/3.0.

Four specimens in the autumn category gave the following results:

1. Specimen 785958, dated Oct. 4, 1964, averaged 4.6Y 3.82/1.8.

16

2. Specimen 369280, dated Sept. 5, 1919, averaged 4.5Y 3.78/1.7.

3. Specimen 369260, dated Aug. 26, 1915, averaged 4.0Y 3.80/1.8.

4. Specimen 369326, dated Sept. 3, 1901, averaged 3.5Y 3.63/2.2.

Some foxing, or discoloration, seemed to have occurred on the oldest of these specimens, and it was therefore omitted from the average. The remaining three specimens averaged 4.4Y 3.8/1.8, which checks fairly well with Hamly's correlation of Ridgway's Olive swatch at 5Y 3.8/3.0. Even if the 1901 specimen were included, the average would be 4.2Y 3.8/2.0—still within the Olive range.

Spectrophotometric testing measured specimen 785958 as 4.6Y 2.8/2.0, very close to the visual average of 4.6Y 3.8/1.8. Specimen 369260 measured 3.2Y 2.9/2.2, compared to the visual average of 4.0Y 3.8/1.8.

After completing the study of the Great Crested Flycatcher, other species of birds described by Ridgway as olive were also investigated. Little by little, as the study was extended, it became increasingly evident that Olive could not be limited to only one single color notation or swatch. Other birds with a very olive appearance matched swatches that were visibly different. A range of such differences emerged, within which Olive was still separable from brownish olive or greenish olive. Some measured close to a 3.3Y 3.95/2.14 notation (approaching brownish olive); others measured 5.1Y 3.80/1.66 (approaching greenish olive). Olive, therefore, must be regarded as within a small range of colors, not merely as one specific color. This "family" concept was found for most of the colors in this guide.

The COLOR GUIDE swatch finally chosen for Olive has a 4.5Y 4.0/2.0 notation. The swatch chosen for Brownish Olive has a 2.5Y 4.0/2.0 notation.

Plan of the Supplement

The Correlated Notes make up the body of the NATURALIST'S COLOR GUIDE SUPPLEMENT. They include a substantial amount of information that is helpful, perhaps even essential, to a complete understanding and interpretation of each color. They have the following general sequence:

1. A discussion of the color and some of the methods and problems that led to its choice and identity in the guide.

2. A tabulation of the color and the colors related to it, including:

    a. The location of comparable color swatches in Ridgway's *Color Standards* and other sources.

    b. Munsell notations of Ridgway's swatches in both *Nomenclature* (1886) and *Color Standards* (1912). Unless otherwise noted, the measurements are those of Hamly. Notations marked with an asterisk are new ones measured for this guide.

c. The Villalobos correlations of Ridgway's colors.

d. Occasional correlations of related colors from Ridgway and from other authors. The correlations do not necessarily agree with each other, nor with the choices made for this guide, but they reflect the best efforts to interpret Ridgway's *Color Standards*.

3. Citations from Bulletin 50 of the species displaying the color under discussion and those related to it. Citations marked with an asterisk are those using the color unequivocally, without modification.

The colors follow each other in a generally logical sequence from reddish colors through the spectrum to bluish and violaceous colors. Some color areas are given more emphasis than others, mainly because the plumage of birds makes this necessary. For example, several rich red and orange-red colors are treated in detail although their hue differences are not substantial. Following the reddish colors are some thirty-six of a brownish or olivaceous nature. This area receives by far the most detailed emphasis for reasons that will be obvious to most naturalists. The series then continues in a more uniform manner through yellow, green, blue, and violet. It ends with grayish colors largely or entirely lacking any hue.

Horn Color is considered in the notes for Raw Umber (Color 23). Rufous and rufescence are discussed in the notes for Cinnamon-Rufous (Color 40).

Appendix A is an alphabetic tabulation of Ridgway's color names and their correlated notations as determined by Ridgway, Hamly, and Villalobos.

Appendix B correlates the nomenclature used by Ridgway in Bulletin 50 and the citations of birds in the Correlated Notes with the nomenclature presently used.

CORRELATED NOTES

Purple ~ Color 1

Purple is a color often thought of as at the violet end of the spectrum. Actually, it does not appear in the spectrum at all. It is composed of a combination of colors from both ends of the spectrum: the high-frequency violets and the low-frequency reds. Because of this peculiarity, and also because it has a reddish quality, Purple is placed at the very beginning of the NATURALIST'S COLOR GUIDE, followed by Magenta, another nonspectral color. Purple should not be confused with violet.

Purple is shown by Ridgway in *Color Standards,* where he gives it a special emphasis, naming it Purple (true) rather than merely Purple. He depicts many other colors ranging from violaceous purples to reddish purples, all of which are omitted from the COLOR GUIDE except Purple and Magenta. One of those omitted, but shown in both Ridgway guides, is Royal Purple. It is not a true purple, however, but has a much darker, nearly violet hue.

Methuen (1967) depicts a purple described as between reddish violet and purplish red. As does Ridgway, he calls it a *true* purple to distinguish it from his nearly violet color called royal purple. He ascribes the color to that of a dye obtained from the marine mollusk Murex, traditionally used to dye the robes of emperors and kings.

Palmer (1962) depicts a swatch called Violet-Magenta. It is close to Ridgway's Purple (true), although it has a slightly redder and decidedly more intense hue.

The COLOR GUIDE swatch chosen for Purple has a 10.0P 4.0/14.0 notation, close to that found for Ridgway's swatch.

Notations of Colors Related to Purple

Color Name	Ridgway Notation	Hamly Notation	Villalobos Notation
Purple (true)	XI 65	10.0 P 4.0/14.0	VM- 8-12°
Purple (true), Hamly alternate		9.5 P 4.2/17.0	
Purple, Guide swatch		10.0 P 4.0/14.0*	
Violet-Magenta, Palmer glossy		0.7 RP 3.4/16.0*	VM- 9-12°
Violet-Magenta, Palmer matte		0.5 RP 3.7/15.5*	
Purple, Methuen		7.0 P 4.0/18.4	
Phlox Purple	XI 65b	10.0 P 5.0/11.0	VM-11-11°

Citations for Purple

No useful citations for Purple were found in Bulletin 50.

Magenta ~ Color 2

Magenta may be pictured as a reddish color with a strong cast of purple. Ridgway depicts it in both *Color Standards* and *Nomenclature.* Palmer (1962) shows it very vividly in his hexagon. It has also been shown in other color guides.

Ridgway's swatch, to judge by the notations Hamly gives it, is much too purple. Palmer's swatch seems to be excessively reddish and vivid, although the notations given by Villalobos for both Ridgway and Palmer are identical (M-10-7°). There is a tendency for Palmer's swatches to show too vivid a color when they are part of his hexagon. The Palmer swatches (both glossy and matte) and the Ridgway swatches (1912 and 1886) were measured spectrophotometrically, using genuinely original copies of Ridgway's guides. The measurements of Ridgway's swatches show a trend toward an acceptable Magenta; Palmer's do not.

A swatch by Methuen (1967) for magenta has a 3.0RP 4.2/14.0 notation. Other possible magenta colors are listed in the tabulation.

Unfortunately, only hummingbirds are described by Ridgway as displaying Magenta. Two of them were measured by spectrophotometer. A Costa Rican Woodstar, called "bright metallic magenta purple," measured 2.3RP 2.5/5.5. A Heloise's Hummingbird, called "brilliant metallic magenta purple," measured 1.7RP 3.2/4.0. It is obvious from these notations that the vivid, iridescent colors were not recorded by the machine, although the hues are not unacceptable.

One of the problems with measurement of iridescent plumage is that the color varies greatly, depending on the angle from which it is viewed. Visual comparisons were therefore made between the plumage and Munsell swatches, which are not available commercially in extremely narrow color ranges, but are close enough to be very helpful. Under correct daylight-lamp conditions, the Costa Rican Woodstar was between 3.75RP 4.0/12.0 and 5.0RP 4.0/12.0; the Heloise's Hummingbird was slightly more purplish than the 3.75RP 4.0/12.0 notation. Naturally, the birds were viewed from angles giving the most purplish red hues, the colors for which I was searching. I believe that these species can be accepted as excellent examples of the color Magenta.

Another test of interest was made on the blossoms of the Amoena Azalea, having a purplish red color that appeared to be Magenta. The blossoms were visually compared with swatches in the Methuen guide and with those of Munsell. The Methuen swatch with the nearest match is 13-C-8, which Methuen equates with a 3.0RP 4.2/14.0 notation. The nearest Munsell swatches ranged from 3.75RP to 5.0RP 4.0/12.0.

The COLOR GUIDE swatch chosen for Magenta has a 3.75RP 4.0/12.0 notation. Magenta might be given a range of hue from 2.0 to 5.0RP, a value close to 4.0/, and a chroma from /10.0 (which is probably too dark) to /14 (which is probably a little too vivid).

Notations of Colors Related to Magenta

Color Name	Ridgway Notation	Hamly Notation	Villalobos Notation
Dull Magenta Purple	XXVI 67'i	10.0 P 4.5/10.5	M- 8-6°
Magenta	XXVI 67'	10.0 P 5.0/12.0	M-10-7°
Magenta, measure of 1912		2.0 RP 4.2/ 9.3*	
Magenta, measure of 1886		1.8 RP 4.8/11.0*	
Magenta, measure of Amoena Azalea		3.75 RP 4.0/12.0*	
Magenta, Guide swatch		3.75 RP 4.0/12.0*	
Magenta, Methuen		3.0 RP 4.2/14.0	
Magenta, Palmer glossy[1]		6.3 RP 3.9/15.5*	M-10-7°
Magenta, Palmer matte		5.5 RP 4.2/14.2*	M-10-7°
Schoenfeld's Purple	XXVI 69'i	5.0 RP 3.2/10.0	MR- 8-7°
Rosolane Purple	XXVI 69'	5.0 RP 4.4/12.0	MR- 9-8°
Permanent Magenta, Winsor & Newton		7.0 RP 3.9/10.8*	

[1]Palmer's Magenta, if based on my measurement of his hexagon, would have a Villalobos notation of M-11-12°, not M-10-7°.

Citations for Magenta

Vol. V   p. 653: chin, throat brilliant metallic solferino or magenta purple (changing to violet)
            ♂Lucifer Hummingbird, *Calothorax lucifer*
*Vol. V   p. 645: gorget bright metallic magenta purple
            ♂Costa Rican Wood-Star, *Nesophlox bryantae*
Vol. V   p. 623: head violet or amethyst purple, changing (certain lights) to violet-blue, even magenta (reddish purple)
            ♂Costa's Hummingbird, *Calypte costae*
*Vol. V   p. 592: chin, throat brilliant metallic magenta purple (changing to other colors)
            ♂Heloise's Hummingbird, *Atthis h. heloisa*
Vol. V   p. 346: throat pomegranate to magenta
            ♂Long-billed Star-throat, *Anthoscenus l. longirostris*

Vinaceous ~ Color 3
Deep Vinaceous ~ Color 4

Vinaceous is a complicated color. Ridgway depicts some forty-five varieties of vinaceous colors in *Color Standards,* using numerous compound names such as Vinaceous-Cinnamon, Russet-Vinaceous, etc. The colors include hues ranging from purplish reds through orange and tones from pale to dark. Ridgway uses these colors generously in Bulletin 50, rarely using the simple name Vinaceous. Apparently, many plumages cannot be described without some reference to their vinaceousness—the precise quality of which is elusive.

The NATURALIST's COLOR GUIDE assumes that the basic quality of the multitude of vinaceous colors is Ridgway's plain Vinaceous—a delicate pink with a slightly purplish cast. The COLOR GUIDE also includes Ridgway's Deep Vinaceous, in which the purplish cast is more visible.

The COLOR GUIDE swatch chosen for Vinaceous has a 2.5R 6.5/7.0 notation; Deep Vinaceous has a 2.5R 5.5/6.0 notation. The tabulation includes a spectrophotometric measurement of Ridgway's Vinaceous swatch, which varies from the COLOR GUIDE swatch, perhaps indicating substantial loss of chromaticity or intensity in the *Color Standards* copy measured.

Notations of Colors Related to Vinaceous

Color Name	Ridgway Notation	Hamly Notation	Villalobos Notation
Pale Vinaceous	XXVII 1''f	2.5 R 7.8/5.0	R-16-6°
Vinaceous	XXVII 1''d	2.5 R 6.5/7.0	R-14-8°
Vinaceous, measure of 1912		10.0 RP 6.4/4.8*?	
Vinaceous, Guide swatch		2.5 R 6.5/7.0*	
Deep Vinaceous	XXVII 1''b	2.5 R 5.6/6.5	RS-12-6°
Deep Vinaceous, measure of 1912		2.5 R 5.7/5.7*	
Deep Vinaceous, Guide swatch		2.5 R 5.5/6.0*	
Dark Vinaceous	XXVII 1''	2.5 R 4.6/5.0	RS-10-5°

Citations for Vinaceous

The following citations are selections of descriptions of pigeons and horned larks, many of which are necessarily of a rather complicated nature. They are listed merely as examples of the usage in Bulletin 50, where many others can be found, none merely Vinaceous.

Vol. VII	p. 446: forehead pale grayish vinaceous to vinaceous-buff, deepening to vinaceous-fawn
	Verreaux's Dove, *Leptotila v. verreauxi*
Vol. VII	p. 363: breast, chest dull russet-vinaceous to light grayish vinaceous, passing to pale grayish vinaceous
	Martinique Dove, *Zenaida aurita*
Vol. VII	p. 326: underparts, ♂vinaceous-drab, passing to vinaceous-buff or vinaceous-fawn; ♀less purplish or vinaceous (nearly army brown or deep fawn color)
	Ruddy Pigeon, *Oenoenas s. subvinacea*
Vol. VII	p. 305: breast, chest purple-drab or vinaceous-drab (area of this color more restricted in ♀)
	Pale-vented Pigeon, *Chloroenas runina pallidicrissa*
Vol. IV	p. 325: hindneck more vinaceous (less rufescent or cinnamomeous)
	Scorched Horned Lark, *Otocoris alpestris adusta*
Vol. IV	p. 320: hindneck less pinkish, more rufescent vinaceous
	California Horned Lark, *Otocoris alpestris actia*
Vol. IV	p. 316: hindneck lighter vinaceous-cinnamon
	Streaked Horned Lark, *Otocoris alpestris strigata*
Vol. IV	p. 311: hindneck lighter, clearer vinaceous
	Prairie Horned Lark, *Otocoris alpestris praticola*
Vol. IV	p. 303: hindneck between dull vinaceous and vinaceous-cinnamon (in spring); vinaceous color concealed in autumn
	Shore Lark, *Otocoris a. alpestris*

Flesh Color ~ Color 5
Salmon Color ~ Color 6

Flesh Color is very rarely used to describe plumage in Bulletin 50; only one acceptable citation was found. Flesh Color is more generally used to describe soft parts, such as the bills and legs of birds. No citations of this sort are given—soft parts quickly lose their "live" coloration, and museum specimens are not likely to be helpful in color identification. However, the color is essential to ornithologists.

Palmer (1962) shows the color, calling it simply Flesh. He identifies it by the notation SO-16-9°, which Villalobos also gives for his measurement of Ridgway's swatch in *Color Standards*. Villalobos gives the same notation to Ridgway's Salmon Color, which indicates that the two swatches in his copy are identical. The same condition is present in my copy. Hamly, on the other hand, must have worked from a substantially different copy; he measures a clear distinction between Flesh and Salmon Color.

Salmon Color, as correlated by Hamly, has a more ochraceous quality than Flesh. In order to arrive at some understanding of Ridgway's Salmon Color, I measured the colors of various species of salmon. King salmon of the Pacific Coast were found to be very vivid in color, measuring from 10.0R 5.0/10.0 to 2.0YR 5.0/11.5. Freshwater landlocked salmon measured nearly cinnamon, lacking pink, with a 6.5YR 6.0/4.0 notation. Salmon from the Atlantic Coast had a notation acceptably close to Hamly's correlation of the Ridgway swatch. It seems reasonable to assume that the flesh of Atlantic Coast salmon was more readily available to Ridgway nearly a hundred years ago, and it was probably the guide to his choice of color.

Three other Ridgway colors occurring in close proximity to Flesh and Salmon in *Color Standards* are recorded at the end of the tabulation. They carry similar notations and perhaps they can be accepted as part of the color range. Flesh Pink is a paler and more clearly pinkish hue than Flesh; Flesh Ocher and Salmon Buff approximately bracket this guide's Salmon Color, however.

The COLOR GUIDE swatch chosen for Flesh Color has a 10.0R 7.0/6.0 notation; it is a somewhat less reddish color than the one shown by Palmer and is ruddier than the swatch for Salmon Color. The COLOR GUIDE swatch chosen for Salmon Color has a 5.0YR 7.0/6.0 notation.

Notations of Colors Related to Flesh and Salmon Colors

Color Name	Ridgway Notation	Hamly Notation	Villalobos Notation
Flesh Color	XIV 7′d	2.5 R 7.0/6.5?	SO-16-9°
Flesh Color, measure of 1912		2.5 YR 6.9/6.4*	
Flesh Color, measure of 1886		1.6 YR 7.6/5.8*	
Flesh, Palmer		10.0 R 7.0/8.0*	SO-16-9°
Flesh Color, Guide swatch		10.0 R 7.0/6.0*	
Salmon Color	XIV 9′d	5.0 YR 7.5/6.0	SO-16-9°
Salmon Color, measure of 1912		4.0 YR 7.1/5.6*	
Salmon Color, measure of 1886		3.5 YR 7.7/6.6*	
Salmon Color, Guide swatch		5.0 YR 7.0/6.0*	

Supplementary Table

Flesh Pink	XIII 5′f	10.0 R 8.0/5.5	S-16-8°
Flesh Ocher	XIV 9′b	2.5 YR 6.6/8.0	SO-15-9°
Salmon Buff	XIV 11′d	7.5 YR 7.8/6.0	SO-17-8°

Citations for Flesh Color

Vol. VI   p. 30: under wing-coverts pale salmon pink or flesh color
Mexican Red-shafted Flicker, *Colaptes c. cafer*

Pink ~ Color 7

Ridgway depicts many pinkish colors, with many different names. They range from purplish through reddish to more orange tones. Two have been related to the Rose "family" in the NATURALIST'S COLOR GUIDE, and three are blended into this Pink "family." Palmer (1962) shows a swatch called Pink; it closely matches the Rose Pink in Ridgway's *Nomenclature*. I rather arbitrarily have chosen to use Palmer's color and color name for Pink in this guide, omitting Ridgway's variations—in fact, he ignores most of them himself.

Two citations from Bulletin 50 were carefully studied. The Pink-headed Warbler, whose head and neck are called "pale rose pink" by Ridgway, shows a wide variation of pinkish tones, depending on the angle from which it is viewed. The plumage is also mixed with delicate silvery feathers. Spectrophotometric measurements were useless for this plumage. Visual comparisons indicated a strong trend to much redder pinks than called for by Ridgway. To say merely "pale pink, shot with silver" is probably as acceptable a description as any. A specimen of Cassin's Purple Finch, whose upper chest is described as "dull rose pink," was so blended with underlying grayish and brownish feathers that the spectrophotometer failed to register the pinkish hues. Visual comparisons indicated that "pale pink with a grayish cast" would be a better description than Rose Pink, which in any event it did not match at all.

The COLOR GUIDE swatch chosen for Pink has a 10.0RP 7.5/6.5 notation.

Notations of Colors Related to Pink

Color Name	Ridgway Notation	Hamly Notation	Villalobos Notation
Thulite Pink	XXVI 71'd	5.0 RP 6.6/11.0	R-16-10°
Thulite Pink, measure of 1912		5.0 RP 6.8/ 6.8*	
Deep Rose Pink	XII 71 d	5.0 RP 6.6/12.0	MR-15-12°
Rose Pink	XII 71 f	5.0 RP 8.0/ 8.0	M-17-12°
Rose Pink, measure of 1912		2.5 RP 7.8/ 6.6*	
Rose Pink, measure of 1886		0.6 R 7.8/ 6.4*	
Pink, Palmer glossy		9.6 RP 7.3/ 6.6*	R-16-11°
Pink, Guide swatch		10.0 RP 7.5/ 6.5*	
Pink, Methuen		6.0 RP 7.3/ 5.6	

Citations for Rose Pink

*Vol. II   p. 760: head, chest rose pink
               ♂Pink-headed Warbler, *Ergaticus versicolor*
*Vol. I   p. 126: rump, below rose pink
               ♂Cassin's Purple Finch, *Carpodacus cassinii*

Carmine ~ Color 8

Carmine is depicted in both of Ridgway's color guides, but Crimson is shown only in *Nomenclature*. Ridgway uses both names throughout Bulletin 50 in about equal degrees. He may have intended some distinction of hue between the two colors, but the various measurements made of his swatches indicate a very close resemblance. Webster's dictionary defines carmine as a rich crimson. Methuen (1967) equates them and describes them as the bluish red color of the organic pigment produced from cochineal. The chromaticity of Methuen's swatch (/14.2) is a little too vivid, but otherwise comes within the range of notations for the colors.

Although omitted from *Color Standards,* Crimson is often used in Bulletin 50; eleven citations are shown, with twelve for Carmine. In order to arrive at an understanding of the colors, I studied the plumages of three species called carmine and four called crimson. One specimen was discarded, primarily due to its iridescent plumage. One specimen showed a trend toward a slightly more orange hue (7.5R 3.0/10.0), but still within the Carmine range of value and chroma. The other five all ranged between 5.0R 3.0/8.0 and 5.0R 3.0/10.0 notations. Ridgway's color name Carmine was chosen for the NATURALIST'S COLOR GUIDE instead of Crimson.

The COLOR GUIDE swatch chosen for Carmine has a 5.0R 3.0/11.0 notation. This more vivid notation is largely influenced by the various color swatches, where the color was more vivid than those of the above-mentioned plumages. Plumages displaying a 3.0/8.0 notation might be called dark carmine, and those with a 3.0/10.0 notation carmine.

Poppy Red requires consideration because it is closely related to the Carmine "family" and because it occurs very frequently in Bulletin 50. Strangely, Ridgway omits it entirely from *Color Standards,* after having clearly depicted it in *Nomenclature.* Measurement of the 1886 swatch indicates a notation that, although paler than Carmine, could not be clearly related to any color of this reddish series other than Carmine. Some thirty citations from Bulletin 50 create the feeling that Ridgway uses poppy red as a basic red against which to compare other colors. About ten refer only to some shade of poppy red, another ten relate it to carmine or crimson, and still another ten to scarlet or vermilion.

Specimens of five species cited as poppy red were examined. Four of them registered between normal carmine and bright carmine. The fifth, the Red, Blue, and Yellow Macaw, called "underparts scarlet vermilion, inclining to deep poppy," was clearly not poppy, but was close to geranium. The other four seem to confirm that poppy red is a bright carmine color:

1. Red-breasted Sapsucker, called a "red nearest poppy," was 5.0R 3.0/10.0 to 7.5R 3.0/10.0.

2. Sapsucker, called "poppy red," was 5.0R 3.0/8.0 to 7.5R 3.0/8.0 with a trend to 3.0/10.0.

3. Pileated Woodpecker, called "bright poppy red," was 7.5R 3.0/10.0 to 7.5R 3.0/12.0.

4. Ant-eating Woodpecker, called "bright poppy red," was 5.0R 3.0/10.0 to 7.5R 3.0/10.0 with a trend to 3.0/8.0.

Notations of Colors Related to Carmine

Color Name	Ridgway Notation	Hamly Notation	Villalobos Notation
Carmine	I 1i	5.0 R 3.4/12.5	RS-7-12°
Carmine, measure of 1912		4.6 R 3.4/10.0*	
Carmine, measure of 1886		3.7 R 4.3/13.3*	RS-7-12°
Carmine, Guide swatch		5.0 R 3.0/11.0*	
Crimson, measure of 1886		4.6 R 3.6/12.0*	RS-6-12°
Crimson, Methuen		4.0 R 4.5/14.2	
Poppy Red, measure of 1886		6.5 R 5.3/11.6*	S-9-12°
Nopal Red	I 3i	6.5 R 4.0/11.0	S-9-12°
Nopal Red, measure of 1912		6.5 R 4.0/10.8*	

Citations for Carmine

Vol. VII	p. 260: small facial area intense poppy red or carmine
	♂Yellow-lored Parrot, *Amazona xantholora*
*Vol. VI	p. 446: throat deep carmine red
	Narrow-billed Tody, *Todus angustirostris*
Vol. VI	p. 347: rump bright poppy red or carmine
	Frantzius' Araçari, *Pteroglossus frantzii*
Vol. VI	p. 342: rump bright poppy red or carmine
	Collared Araçari, *Pteroglossus t. torquatus*
*Vol. VI	p. 317: pileum nearly carmine; throat paler, more poppy red
	♂Salvin's Barbet, *Eubocco bourcieri salvini*
*Vol. VI	p. 181: pileum, crest, nape, chest, etc., carmine red
	♂Splendid Woodpecker, *Cniparchus haematogaster splendens*
Vol. VI	p. 89: crown between poppy red and carmine
	♂Golden-cheeked Woodpecker, *Centurus c. chrysogenys*
Vol. VI	p. 71: crown, nape between poppy red and carmine
	♂Swainson's Woodpecker, *Centurus r. rubriventris*
Vol. V	p. 666: pileum intense rose red to carmine
	♂Ruby-and-Topaz Hummingbird, *Chrysolampis mosquitus*
Vol. II	p. 760: forehead, rump shades of burnt carmine; head, chest rose pink
	♂Pink-headed Warbler, *Ergaticus versicolor*

31

| Vol. II | p. 759: rump, below between poppy red and carmine
♂Red Warbler, *Ergaticus ruber* |
| Vol. I | p. 53: head, rump, below usually light carmine or pinkish red
♂White-winged Crossbill, *Loxia leucoptera* |

Citations for Crimson

Vol. VII	p. 125: back dull crimson, below nearly carmine Red, Blue, and Green Macaw, *Ara chloroptera*
Vol. VI	p. 174: head, crest poppy red to crimson ♂Guatemalan Ivory-billed Woodpecker, *Scapaneus g. guatemalensis*
*Vol. VI	p. 131: pileum, nape, etc., bright crimson ♂Yucatan Woodpecker, *Chloronerpes rubiginosus yucatanensis*
*Vol. VI	p. 129: nape, etc., bright crimson ♂Lichtenstein's Woodpecker, *Chloromerpes aeruginosus*
*Vol. VI	p. 42: head, neck, upper chest bright crimson Red-headed Woodpecker, *Melanerpes erythrocephalus*
Vol. V	p. 605: gorget between bright metallic crimson or burnt carmine and pomegranate purple ♂Glow-throated Hummingbird, *Selasphorus ardens*
Vol. V	p. 489: gorget purplish red, varying from solferino to nearly crimson ♂Garnet-throated Hummingbird, *Lamprolaima rhami*
Vol. II	p. 116: above dark crimson-maroon, below, blood-red ♂Crimson-backed Tanager, *Ramphocelus d. dimidiatus*
Vol. I	p. 131: face, forehead, throat, rump varying crimson to poppy red ♂Mexican House Finch, *Carpodacus m. mexicanus*
Vol. I	p. 128: pileum deep wine-purple, more crimson in summer ♂Purple Finch, *Carpodacus p. purpureus*
Vol. I	p. 126: pileum dull crimson, brighter in summer ♂Cassin's Purple Finch, *Carpodacus cassinii*

Citations for Poppy Red (1886)

Vol. VII	p. 260: small facial area intense poppy red or carmine ♂Yellow-lored Parrot, *Amazona xantholora*
Vol. VII	p. 242: forehead, crown between poppy red and geranium Red-crowned Parrot, *Amazona viridigenalis*
*Vol. VII	p. 152: forehead, part of crown deep poppy red Finch's Paroquet, *Aratinga finchi*
*Vol. VII	p. 132: forehead, lores bright poppy red Mexican Green Macaw, *Ara militaris mexicana*
Vol. VII	p. 128: underparts inclining to deep poppy red Red, Blue and Yellow Macaw, *Ara Macao*
Vol. VII	p. 125: head, neck bright poppy red, back dull crimson Red, Blue, and Green Macaw, *Ara chloroptera*
Vol. VI	p. 347: rump bright poppy red or carmine Frantzius' Araçari, *Pteroglossus frantzii*
Vol. VI	p. 342: rump bright poppy red or carmine Collared Araçari, *Pteroglossus t. torquatus*

Vol. VI	p. 317: pileum nearly carmine, throat paler, more poppy red ♂Salvin's Barbet, *Eubocco bourcieri salvini*
*Vol. VI	p. 282: head, neck, chest bright red, nearest poppy red (in spring) ♂Red-breasted Sapsucker, *Sphyrapicus r. ruber*
*Vol. VI	p. 274: chin, throat poppy red, crown may be crimson ♂Sapsucker, *Sphyrapicas v. varius*
Vol. VI	p. 233: nuchal band red (not defined) ♂Southern Downy Woodpecker, *Dryobates p. pubescens*
Vol. VI	p. 210: nuchal band bright poppy red or scarlet ♂Hairy Woodpecker, *Dryobates v. villosus*
Vol. VI	p. 174: head, crest poppy red to crimson ♂Guatemalan Ivory-billed Woodpecker, *Scapaneus g. guatemalensis*
*Vol. VI	p. 155: pileum, crest bright poppy red ♂Pileated Woodpecker, *Phloeotomus p. pileatus*
*Vol. VI	p. 102: crown, nape bright poppy red ♂Ant-eating Woodpecker, *Balanosphyra f. formicivora*
Vol. VI	p. 89: crown between poppy red and carmine ♂Golden-cheeked Woodpecker, *Centurus c. chrysogenys*
Vol. VI	p. 71: crown, nape between poppy red and carmine ♂Swainson's Woodpecker, *Centurus r. rubriventris*
Vol. IV	p. 747: head bright vermilion to poppy red ♂Southern Yellow-thighed Manakin, *Pipra mentalis ignifera*
Vol. IV	p. 475: pileum intense poppy red, scarlet-vermilion, or scarlet; below similar but lighter, more orange ♂Vermilion Flycatcher, *Pyrocephalus rubinus mexicanus*
Vol. II	p. 759: rump and below between poppy red and carmine ♂Red Warbler, *Ergaticus ruber*
Vol. II	p. 728: breast, belly poppy red or rich vermilion red Painted Redstart, *Setophaga p. picta*
Vol. II	p. 720: face, chin, throat vermilion or poppy red ♂Red-faced Warbler, *Cardellina rubrifrons*
Vol. II	p. 330: wing shoulder bright poppy red or vermilion, to scarlet or even (rarely) orange-chrome (in summer) ♂Red-winged Blackbird, *Agelaius p. phoeniceus*
Vol. II	p. 326: wing-coverts rich poppy red or vermilion ♂Bicolored Blackbird, *Agelaius g. gubernator*
Vol. II	p. 99: general color bright vermilion or poppy red, some inclining to scarlet ♂White-winged Tanager, *Piranga l. leucoptera*
*Vol. II	p. 79: dark dull poppy red, rump clearer, underparts rich vermilion or poppy red (in summer) ♂Summer Tanager, *Piranga r. rubra*
*Vol. I	p. 85: crown bright poppy red ♂Redpoll, *Acanthis l. linaria*
Vol. I	p. 80: crown bright vermilion or poppy red (in spring) ♂Greenland Redpoll, *Acanthis h. hornemannii*
Vol. I	p. 60: head, below general color light poppy red (in summer), dull pinkish red (in winter) ♂Canadian Pine Grosbeak, *Pinicola enucleater canadensis*

Rose ~ Color 9

Rose Color shown by Ridgway is practically equivalent to Rose shown by Palmer (1962). Palmer's swatch is acceptable for the color, and the NATURALIST's COLOR GUIDE uses the shorter name. Another Ridgway color, Rose Red, is called Ruby in this guide (Color 10), which eliminates the confusion that can result from using Ridgway's Rose Color and Rose Red as color names.

The tabulation indicates a rather wide range of hues, especially when Hamly's 2.5RP notation for Tyrian Pink is included. It was difficult to make clear-cut distinctions between the numerous pink colors, various rose colors, and related rose red colors, not to mention other reddish colors. However, the Villalobos notations were helpful; they equate all the colors listed for the Rose range. I also studied the Ridgway swatches, which were in close proximity to each other in both *Color Standards* and *Nomenclature,* and measured many of them spectrophotometrically. I have concluded that Palmer's color name and swatches are preferable to Ridgway's in this instance.

The COLOR GUIDE swatch chosen for Rose has a 6.5RP 5.0/14.0 notation.

Notations of Colors Related to Rose

Color Name	Ridgway Notation	Hamly Notation	Villalobos Notation
Rose Color	XII 71b	5.0 RP 5.5/14.0	MR-12-11°
Rose Color, measure of 1912		6.3 RP 5.3/12.6*	
Spinel Pink	XXVI 71'b	5.0 RP 5.2/12.0	MR-13-11°
Tyrian Pink	XII 69b	2.5 RP 5.4/12.0?	MR-12-11°
Tyrian Pink, measure of 1912		4.2 RP 5.3/12.6*	
Rose, Palmer glossy		6.4 RP 5.1/13.4*	MR-12-11°
Rose, Palmer matte		6.4 RP 5.4/13.0*	MR-12-11°
Rose, Guide swatch		6.5 RP 5.0/14.0*	

Citations for Rose

No useful citations for Rose were found in Bulletin 50.

Ruby ~ Color 10

Ridgway's Rose Red is a more vivid red than his Rose Color, but still not a clearly unequivocal red color. As is the case with the paler Rose Color, Rose Red has a wide range of hue, from about 8.5RP 5.0/ 12.0 to 2.7R 4.5/14.0. The notations given by Villalobos, on the other hand, are nearly identical and indicate a very narrow range.

Included within this color range is a swatch shown by Palmer (1962) called Ruby, with a 2.6R 4.6/15.0 notation in a matte condition. It is one of the intense colors of his hexagon, as it is of the Villalobos hexagon. The use of the name Ruby for the reddish colors of this area can be questioned. Ruby gemstones are not usually the color of the Palmer swatch for Ruby, nor are they the color of any of the swatches in the Rose Red group. The gem ruby is most often darker, perhaps more carmine or even a bluish red shade. However, Ridgway's names Rose Color and Rose Red can only result in confusion. Although I am reluctant to differ so materially from Ridgway—and also from the color of some ruby gems—I have decided to follow Palmer and substitute the name Ruby for Ridgway's Rose Red.

Ridgway uses Rose Red in both his guides, but his use of the color in Bulletin 50 is not extensive. One citation for ruby is listed, although he does not depict it with a swatch.

A specimen of the Mexican Thrush-Warbler, described as "clear rose red," was measured by spectrophotometer. It registered a 5.4R 4.4/10.5 notation. Visual comparisons indicated close to a 2.5R 5.0/ 15.0 notation. A specimen of the Panama Thrush-Warbler, also described as "clear rose red," measured 4.4R 3.5/8.7, with a visual notation of about 2.5R 5.0/12.0. A specimen of the Ruby-and-Topaz Hummingbird is described variably by Ridgway as "rose red to carmine," "intense metallic red," and "brilliant metallic ruby." The spectrophotometric measurement of the iridescent plumage of this specimen could not be used, but visual comparisons registered 5.0R 4.0/14.0 and 7.5R 4.0/14.0 notations.

The tests outlined above indicate a substantial loss of chroma when some plumages are measured by spectrophotometer, as compared with the greater chromaticity found in visual comparisons. We can accept the two Thrush-Warblers as Ruby, but the Ruby-and-Topaz Hummingbird is better related to Spectrum Red (Color 11) and Scarlet (Color 14).

The COLOR GUIDE swatch chosen for Ruby has a 2.5R 5.0/14.0 notation.

Notations of Colors Related to Ruby

Color Name	Ridgway Notation	Hamly Notation	Villalobos Notation
Rose Red	XII 71	7.5 RP 4.8/14.0	R-10-12°
Rose Red, measure of 1912		2.7 R 4.4/13.5*	
Tyrian Rose	XII 69	5.0 RP 4.5/14.0	R- 9-12°
Tyrian Rose, Hamly alternate		8.0 RP 4.3/15.0	R- 9-12°
Permanent Rose, Winsor & Newton		10.0 RP 5.2/14.2*	
Spinel Red	XXVI 71	8.5 RP 5.0/12.0	R-10-10°
Spinel Red, measure of 1912		10.0 RP 4.5/10.6*	
Ruby, Palmer glossy		3.1 R 4.4/14.5*	R-10-12°
Ruby, Palmer matte		2.6 R 4.6/15.0*	R-10-12°
Ruby, Guide swatch		2.5 R 5.0/14.0*	

Citations for Rose Red

*Vol. V   p. 666: pileum intense metallic red, rose red to carmine
       ♂Ruby-and-Topaz Hummingbird, *Chrysolampis mosquitus*

*Vol. V   p. 625: head brilliant metallic rose red (pileum more purplish)
       ♂Helena's Hummingbird, *Calypte helenae*

Vol. V   p. 619: head largely brilliant metallic rose red, changing (certain lights) to solferino and violet
       ♂Anna Hummingbird, *Calypte anna*

Vol. V   p. 616: crown, chin, throat glittering rose red, changing (certain lights) to scarlet, especially on chin and throat
       ♂Floresi's Hummingbird, *Selasphorus floresii*

Vol. V   p. 350: throat bright metallic red, varying rose red to scarlet (only fresh plumage)
       ♂Constant's Star-throat, *Anthoscenus c. constantii*

*Vol. II   p. 772: eyestripe, face, below rose red
       ♂Mexican Thrush-Warbler, *Rhodinocichla schistacea*

*Vol. II   p. 771: eyestripe, face, below rose red
       ♂Panama Thrush-Warbler, *Rhodinocichla rosea eximia*

Vol. I   p. 614: chest, midbreast rose red or light carmine (and other variations)
       ♂Rose-breasted Grosbeak, *Zamelodia ludoviciana*

Citations for Ruby

Vol. V   p. 665, pileum brilliant ruby red (from rose red to carmine)
       ♂Ruby-and-Topaz Hummingbird, *Chrysolampis mosquitus*

Spectrum Red ~ Color 11

Spectrum Red and Scarlet-Red are both shown in *Color Standards,* where they are nearly identical. The name Scarlet-Red can be too easily confused with Ridgway's Scarlet, which is an entirely different color with a strongly orange tinge (Color 14). Only if the name Scarlet-Red were invariably used in its hyphenated form could confusion with the name Scarlet be avoided. The NATURALIST'S COLOR GUIDE therefore omits it as a color name in favor of Spectrum Red.

This is the first example of the use of *spectrum* to designate a specific color. It is used in the COLOR GUIDE for each of the colors of the spectrum when the color occurs in its purest form. This procedure should eliminate the confusion when a color is called red or green or blue, meaning perhaps a color or perhaps merely reddish, greenish, or bluish. Spectrum Red is the purest red available, with a minimum (if any) tinge of purplish hue at one extreme or orange at the other.

Methuen (1967) describes Spectrum Red as the color of the rays of sunlight having a wavelength of 614 millimicrons (mμ), and also with an approximate range of 4.0R 4.0/12.0 to 7.0R 4.8/15.5 notations. These two extremes are rather arbitrarily averaged to 5.5R 4.7/15.0, which is close to the notation of the COLOR GUIDE swatch.

Regrettably, Ridgway rarely uses Spectrum Red or Scarlet-Red in Bulletin 50. In fact, I found possibly useful citations for Spectrum Red only in small, subordinate areas of plumage and none at all for Scarlet-Red. The Cuban Parrot, which he describes as spectrum red in a small area of the rectrices, was more nearly carmine and could not be used to confirm his description. However, I found that the Ruby-and-Topaz Hummingbird was acceptably close, although called rose red to carmine by Ridgway. When examined with the head held downward toward the viewer, a specimen registered a 5.0R 4.0/14.0 notation, well within the color range of Spectrum Red. However, when examined with head upward, the iridescence became more scarlet.

The COLOR GUIDE swatch chosen for Spectrum Red has a 5.0R 4.0/15.0 notation.

Notations of Colors Related to Spectrum Red

Color Name	Ridgway Notation	Hamly Notation	Villalobos Notation
Spectrum Red	I 1	5.0 R 4.0/15.0	RS- 9-12°
Spectrum Red, measure of 1912		5.8 R 4.3/13.5*	
Scarlet Red	I 3	5.0 R 4.4/16.0	RS-10-12°
Scarlet Red, measure of 1912		6.7 R 4.5/14.0*	
Spectrum Red, Guide swatch		5.0 R 4.0/15.0*	
Spectrum Red, average of Methuen		5.5 R 4.7/15.0	

Citations for Scarlet-Red

No useful citations for Scarlet-Red were found in Bulletin 50.

Citations for Spectrum Red

*Vol. VII   p. 270: small area of rectrices spectrum red
             Cuban Parrot, *Amazona l. leucocephala*
Vol. VII   p. 266: area of rectrices dull spectrum red or chinese vermilion
             Santo Domingo Parrot, *Amazona ventrilis*
Vol. VII   p. 262: inner webs, four outer rectrices basally dull spectrum red
             Lesser Jamaican Parrot, *Amazona agilis*
*Vol. VII   p. 260: bend of wing, primary coverts pure spectrum red
             Yellow-lored Parrot, *Amazona xantholora*
Vol. VII   p. 239: outer webs, three secondaries poppy red or dull spectrum red
             Plain-colored Parrot, *Amazona f. farinosa*
Vol. VII   p. 210: proximal two-thirds of inner webs, some rectrices dull spectrum red or chinese vermilion
             Blue-headed Parrot, *Pionus menstruus*
Vol. VII   p. 200: parts of wing-coverts dull spectrum red or pure chinese vermilion
             Red-fronted Parrotlet, *Urochroma costaricensis*

Geranium ~ Color 12

Geranium is a color with a slightly more orange hue than Spectrum Red and less orange than Scarlet. It is depicted as Geranium Red in *Nomenclature,* but not in *Color Standards.* Ridgway's descriptions in Bulletin 50 relate primarily to certain trogons, whose underparts are specifically called "pure geranium red." Trogon plumage of these areas fades out materially, more rapidly than is the case for most other birds. It was essential to measure fresh plumage of four species of trogons to find Ridgway's color. I found it to be somewhat more orange in quality than the color indicated by measurement of the 1886 swatch.

The Ridgway name has been reduced here simply to Geranium. The color may be typified by the flowers of the same name, particularly the more deeply colored types. Measurements of geraniums varied widely but centered about a 7.5R 4.0/15.0 notation, which is the color chosen for the COLOR GUIDE swatch for Geranium.

Notations of Colors Related to Geranium

Color Name	Ridgway Notation	Hamly Notation	Villalobos Notation
Geranium Red, measure of 1886		5.0 R 5.5/14.7*?	RS-11-12°
Grenadine Red	II 7	8.5 R 5.4/13.0	SSO-12-12°
Geranium, Guide swatch		7.5 R 4.0/15.0*	
Grenadine	II 7b	9.0 R 6.0/12.0	SSO-15-12°

Citations for Geranium Red

Vol. V	p. 776: below pure deep geranium red
	Jalapa Trogon, *Trogonurus puella*
*Vol. V	p. 773: below pure geranium red
	♂Elegant Trogon, *Trogonurus elegans*
*Vol. V	p. 768: below pure geranium red
	♂Coppery-tailed Trogon, *Trogonurus a. ambiguus*
*Vol. V	p. 765: below pure geranium red
	♂Mexican Trogon, *Trogonurus mexicanus*
*Vol. V	p. 749: below pure geranium red
	Lattice-tailed Trogon, *Curucujus clathratus*
*Vol. V	p. 747: below pure geranium red
	Large-tailed Trogon, *Curucujus melanurus macrourus*
*Vol. V	p. 744: below pure geranium red
	Massena Trogon, *Curucujus massena*

Vol. V   p. 741: below bright geranium red
           ♂Eared Trogon, *Leptuas neoxenus*
Vol. V   p. 736: chest intense geranium red, darkening to crimson or
           burnt carmine on upper breast
           ♂Quetzal, *Pharomachrus m. mocinno*
Vol. V   p. 629: throat nearest geranium red, in certain lights
           ♂Ruby-throated Hummingbird, *Archilochus colubris*
Vol. II   p. 701: below vermilion or geranium red
           ♂Salle's Red-breasted Chat, *Granatellus s. sallaei*

Geranium Pink ~ Color 13

Geranium Pink is depicted by Ridgway in *Color Standards*. It is a pale variety of Geranium, perhaps similar to the faded plumage of the birds considered for Color 12. Here again, I found that the Ridgway swatch lacked the more orange hue of the birds and also that of the paler geranium flowers. There appears to have been some alteration of the colors in this portion of *Color Standards*. My choice of color is closer to Ridgway's Rose Doree, at least as it appears in my copy of *Color Standards*.

The COLOR GUIDE swatch chosen for Geranium Pink has a 6.25R 5.0/14.0 notation.

Notations of Colors Related to Geranium Pink

Color Name	Ridgway Notation	Hamly Notation	Villalobos Notation
Geranium Pink	I 3d	4.0 R 6.5/11.0	RS-15-12°
Geranium Pink, measure of 1912		3.8 R 6.5/ 9.7*	
Geranium Pink, Guide swatch		6.25 R 5.0/14.0*	
Begonia Rose	I 1b	5.0 R 5.6/11.5	RS-13-12°
Rose Doree	I 3b	6.0 R 6.0/12.0	RS-13-12°

Citations for Geranium Pink

No useful citations for Geranium Pink were found in Bulletin 50.

Scarlet ~ Color 14

Scarlet, Scarlet Vermilion, and Vermilion are equated with each other in the NATURALIST'S COLOR GUIDE. Ridgway depicts Scarlet in *Color Standards* and *Nomenclature;* Scarlet Vermilion and Vermilion only in *Nomenclature.* The tabulation of notations of these colors indicates a very close, nearly identical relationship. The notations call for a hue leaning toward orange and a very vivid chroma or intensity. They indicate a characteristic distinction from Ridgway's Scarlet-Red, equated with Spectrum Red (Color 11) in this guide. Inasmuch as Scarlet-Red has been discarded, it should not be confusing to call this color Scarlet, though some may picture it as a purer red color.

Palmer (1962) includes a color called Scarlet. When measured spectrophotometrically, it proved nearly equivalent to Ridgway's Scarlet and Vermilion. If we use the Villalobos notation (SSO-10-12°), it is identical to Ridgway's Scarlet; if we use the Villalobos hexagon of colors (S-10-12°), it is identical to Ridgway's Vermilion.

Methuen (1967) equates scarlet not only with vermilion but also with a selected tone of cinnabar, all three carrying a cast of reddishness. Methuen says: "Scarlet, same as vermilion, and cinnabar. Cinnabar is an archaic name for scarlet and vermilion. The pigment cinnabar varies in color." However, Methuen gives cinnabar the typical notation of 8.0R 5.3/15.4.

Ten species of birds described by Ridgway as having a scarlet or vermilion plumage were measured by direct comparison with Munsell swatches. The color of Passerini's and Cherrie's Tanagers proved to be the best examples of Scarlet. Half of the ten species would seem better placed in the Geranium range of colors.

Du Bus' Red-breasted Chat, called "pure vermilion red," measured between 6.25R 4.0/14.0 and 7.5R 4.0/14.0, trending to 5.0/14.0. This plumage is close to that of the Massena Trogon shown as Geranium (Color 12). Ridgway calls the Salle's Chat "vermilion or geranium red."

Underparts of the Summer Tanager are called "clear rich vermilion or poppy red." The specimen matched fairly closely to the Massena Trogon; it might better be called Geranium. It measured 7.5R 4.0/14.0, trending toward an 8.75 hue.

Cardinal underparts are called "pure vermilion red." Three specimens were tested; they varied only slightly from one another. All ranged closer to Geranium than to Scarlet, with 7.5R 4.0/12.0 and 7.5R 5.0/14.0 notations.

A Flicker's nuchal area, called "bright scarlet," had a 6.25R 3.0/ 12.0 notation, which is better placed in the bright Carmine range.

A Painted Bunting, called "vermilion red," is better placed as a dull Geranium color, having a notation of 7.5R 4.0/14.0.

The underparts of a Vermilion Flycatcher are described as similar to the color of the pileum and called "intense poppy, scarlet-vermilion, or scarlet," but "varying toward orange." We accept this plumage as Scarlet, with 8.75R 5.0/14.0 to 8.75R 4.5/16.0 notations.

Passerini's Tanager, called "intense scarlet," and Cherrie's Tanager, called "bright, pure, intense scarlet," were most clearly Scarlet, with 7.5R 4.0/15.5 and 8.75R 4.5/16.5 notations and a color close to Palmer's Scarlet swatch.

A Scarlet Tanager is called "intense scarlet . . . the red varying to almost orange hues." The crown measured nearer Geranium, with a 7.5R 4.0/14.0 notation, but upper parts were acceptably Scarlet, with an 8.75R 4.5/16.5 notation, underparts slightly more orange.

A Macaw (*Ara macao*), with underparts called "deep scarlet-vermilion, inclining to poppy," matched fairly closely to the Geranium plumage of the Large-tailed Trogon, possibly with a little richer reddish hue or a deep Geranium color. Notations varied somewhat for different body areas: approximately 8.75R 4.0/14.0 to 3.0/12.0 and 7.5R 4.0/14.0.

The COLOR GUIDE swatch chosen for Scarlet has an 8.75R 4.5/16.5 notation, very close to that published by Palmer. It has been chosen to depict a trend toward orange and a very vivid chromaticity.

Notations of Colors Related to Scarlet

Color Name	Ridgway Notation	Hamly Notation	Villalobos Notation
Scarlet	I 5	7.5 R 5.0/12.0	SSO-10-12°
Scarlet, measure of 1912		9.5 R 5.1/12.6*	
Scarlet, measure of 1886		8.0 R 6.0/14.3*	SSO-10-12°
Scarlet, Palmer glossy		8.2 R 4.3/17.0*	SSO-10-12°
Scarlet = Vermilion = Cinnabar, Methuen		8.0 R 5.3/15.4	
Scarlet = Vermilion, 1886			S-10-10°
Vermilion, measure of 1886		7.5 R 5.3/13.0*	S-10-11°
Scarlet, Guide swatch		8.75 R 4.5/16.5*	
Cinnabar, Methuen		8.0 R 5.3/15.4	

Citations for Vermilion (1886)

Vol. VII   p. 234: forehead broadly dull vermilion or bright nopal red
Yellow-cheeked Parrot, *Amazona a. autumnalis*

Vol. VII	p. 200: frontal half crown dull scarlet or chinese vermilion ♂Red-fronted Parrotlet *Urochroma costaricensis*
*Vol. VII	p. 136: forehead broadly vermilion red Cuban Macaw, *Ara tricolor*
Vol. IV	p. 747: head bright vermilion to poppy red ♂Southern Yellow-thighed Manakin, *Pipra mentalis ignifera*
*Vol. III	p. 705: crownpatch vermilion red (rarely salmon) (in spring) ♂Ruby-crowned Kinglet, *Regulus c. calendula*
*Vol. II	p. 732: below vermilion red ♂Red-bellied Redstart, *Myioborus m. miniatus*
Vol. II	p. 728: breast, belly rich vermilion red or poppy red Painted Redstart, *Setophaga p. picta*
Vol. II	p. 720: face, chin, throat vermilion or poppy red ♂Red-faced Warbler, *Cardellina rubrifrons*
Vol. II	p. 701: below pure vermilion red or geranium red ♂Salle's Red-breasted Chat, *Granatellus s. sallaei*
*Vol. II	p. 699: below pure vermilion red ♂DuBus' Red-breasted Chat, *Granatellus venustus*
Vol. II	p. 330: wing-coverts bright poppy red or vermilion (to scarlet or even rarely orange-chrome) (in summer) Red-winged Blackbird, *Agelaius p. phoeniceus*
Vol. II	p. 326: wing-coverts rich poppy red or vermilion ♂Bicolored Blackbird, *Agelaius g. gubernator*
*Vol. II	p. 151: throat, chest light vermilion red ♂Yucatan Ant Tanager, *Phoenicothraupis salvini peninsularis*
*Vol. II	p. 148: throat only vermilion red ♂Salvin's Ant Tanager, *Phoenicothraupis s. salvini*
Vol. II	p. 102: pileum vermilion or orange-vermilion; face, throat pale vermilion or pinkish red ♂Red-headed Tanager, *Piranga erythrocephala*
Vol. II	p. 99: general color bright vermilion or poppy red, some inclining to scarlet ♂White-winged Tanager, *Piranga l. leucoptera*
*Vol. II	p. 83: above dull vermilion (rump clearer); below light vermilion ♂Western Summer Tanager, *Piranga rubra cooperi*
Vol. II	p. 79: above dark dull poppy red (rump clearer); below rich vermilion or poppy red (in summer) ♂Summer Tanager, *Piranga r. rubra*
*Vol. I	p. 642: head, neck, below intense pure vermilion ♂Jalapa Cardinal, *C. cardinalis coccineus*
*Vol. I	p. 635: head, below pure vermilion (back duller) ♂Cardinal, *C. c. cardinalis*
*Vol. I	p. 586: below, including throat, vermilion red ♂Painted Bunting, *Cyanospiza ciris*
Vol. I	p. 80: crown bright vermilion or poppy red (in spring) ♂Greenland Redpoll, *Acanthis h. hornemannii*

44

Citations for Scarlet

Vol. VII   p. 200: frontal half crown dull scarlet or chinese vermilion
         ♂Red-fronted Parrotlet, *Urochroma costaricensis*

Vol. VI   p. 201: nuchal band bright poppy red or scarlet
         ♂Hairy Woodpecker, *Dryobates v. villosus*

*Vol. VI   p. 14: nuchal band bright scarlet
         Flicker, *Colaptes a. auratus*

*Vol. V   p. 616: chin, throat scarlet in certain lights
         Floresi's Hummingbird, *Selasphorus floresii*

Vol. V   p. 612: chin, throat brilliant metallic scarlet (changing to golden green in certain lights)
         ♂Rufous Hummingbird, *Selasphorus rufus*

Vol. V   p. 610: chin, throat brilliant scarlet or orange-red (changing to to golden and greenish in certain lights)
         ♂Allen's Hummingbird, *Selasphorus alleni*

Vol. IV   p. 475: pileum intense poppy red, scarlet-vermilion, or scarlet, underparts similar but lighter, toward orange
         ♂Vermilion Flycatcher, *Pirocephalus rubinus mexicanus*

Vol. II   p. 330: wing-coverts scarlet and other colors
         ♂Red-winged Blackbird, *Agelaius p. phoeniceus*

*Vol. II   p. 144: crest scarlet
         ♂Mexican Ant Tanager, *Phoenicothraupis rubica rubicoides*

Vol. II   p. 115: lower back, rump, below glossy scarlet or scarlet vermilion
         ♂Bonaparte's Tanager, *Ramphocelus luciani*

*Vol. II   p. 111: lower back, rump intense scarlet
         ♂Cherrie's Tanager, *Ramphocelus costaricensis*

*Vol. II   p. 109: lower back, rump intense scarlet
         ♂Passerini's Tanager, *Ramphocelus passerinii*

Vol. II   p. 88: all but wings and tail intense scarlet or scarlet-vermilion (in spring)
         ♂Scarlet Tanager, *Piranga erythromelas*

Citations for Scarlet Vermilion

See citations for Scarlet (Vol. II, pp. 88, 115; Vol. IV, p. 475).

*Vol. VII   p. 134: lores, forehead bright scarlet-vermilion
         Buffon's Macaw, *Ara ambigua*

*Vol. VII   p. 128: head, neck, tail (largely) scarlet vermilion; underparts deeper, inclining to deep poppy red
         Red, Blue, and Yellow Macaw, *Ara macao*

Flame Scarlet ~ Color 15

Flame Scarlet is shown in *Color Standards,* but is rarely used in Bulletin 50. Undoubtedly there are plumages carrying this distinctive color, more orange than Scarlet, but more fiery than Chrome Orange. Palmer (1962) depicts a swatch called Scarlet Orange; it is nearly identical to Flame Scarlet, both visually and spectrophotometrically. Villalobos equates Ridgway's Orange Chrome with Flame Scarlet.

The COLOR GUIDE swatch chosen for Flame Scarlet has a 10.0R 5.0/16.0 notation.

Notations of Colors Related to Flame Scarlet

Color Name	Ridgway Notation	Hamly Notation	Villalobos Notation
Flame Scarlet	II 9	10.0 R 6.0/14.0	SO-13-12°
Flame Scarlet, measure of 1912		0.8 YR 5.6/12.2*	
Flame Scarlet, measure of 1886		10.0 R 6.1/13.1*	
Flame Scarlet, Guide swatch		10.0 R 5.0/16.0*	
Scarlet Orange, Palmer glossy		0.2 YR 5.5/16.0*	
Scarlet Orange, Palmer matte		10.0 R 5.3/14.1*	
Saturn Red, 1886			SO-14-12°

Citations for Flame Scarlet

Vol. V   p. 753: belly pure reddish orange or orange-red (flame scarlet, to between flame scarlet and saturn red); ♀orange chrome to saturn red to light flame scarlet
♂Baird's Trogon, *Trogon bairdii*

Vol. IV   p. 746: head intense flame scarlet or orange-vermilion
♂Yellow-thighed Manakin, *Pipra m. mentalis*

Vol. II   p. 311: some coverts and below varying cadmium yellow to intense orange or almost flame scarlet (average hue cadmium orange) (in summer)
♂Baltimore Oriole, *Icterus galbula*

Chrome Orange ~ Color 16

Ridgway's Orange Chrome approaches Spectrum Orange, but still retains some reddish quality—less than Flame Scarlet, however. It is represented by a swatch in *Color Standards,* but is not greatly used by Ridgway in Bulletin 50. Six of his citations are somewhat confusingly related to more reddish tones, particularly to Orpiment Orange, which is shown only in *Nomenclature.* It is possible that some of his citations are closer to Flame Scarlet colors.

Ridgway's color name is altered in the NATURALIST's COLOR GUIDE to Chrome Orange; orange is the basic hue, not chrome.

The COLOR GUIDE swatch chosen for Chrome Orange has a 2.5YR 6.0/16.0 notation, which is close to that given by Hamly for Ridgway's swatch.

Notations of Colors Related to Chrome Orange

Color Name	Ridgway Notation	Hamly Notation	Villalobos Notation
Orange Chrome	II 11	2.5 YR 6.0/15.0	SO-13-12°
Orange Chrome, measure of 1912		2.8 YR 6.0/13.5*	
Orange Chrome, measure of 1886		2.8 YR 6.2/12.7*	
Chrome Orange, Guide swatch		2.5 YR 6.0/16.0*	
Chrome Orange, Winsor & Newton		1.4 YR 6.0/13.0*	
Orpiment Orange, 1886			SO-12-12°

Citations for Orange Chrome

The orange red of these citations is not a Ridgway color, but is merely a descriptive term.

Vol. IV p. 711: crown patch bright orange or orange red (orpiment orange to orange chrome)
♂Giant Kingbird, *Tyrannus cubensis*

Vol. IV p. 706: crown patch orange or orange red (orpiment orange to orange chrome)
♂Gray Kingbird, *Tyrannus d. dominicensis*

*Vol. IV p. 700: crown patch bright reddish orange (orange chrome)
♂Lichtenstein's Kingbird, *Tyrannus melancholicus satrapa*

Vol. IV p. 697: crown patch reddish orange or orange red (orpiment orange to orange chrome)
♂Arkansas Kingbird, *Tyrannus verticalis*

*Vol. IV p. 694: crown patch orange red (orange chrome)
♂Cassin's Kingbird, *Tyrannus vociferans*

Vol. II p. 95: pileum near orange chrome
♂Swainson's Tanager, *Piranga b. bidentata*

47

Spectrum Orange ~ Color 17

Spectrum Orange is the name chosen in accordance with the plan to use *spectrum* with each of the six colors of the spectrum. Using Spectrum Orange eliminates Ridgway's names Cadmium Orange and Orange, shown in both his color guides. Cadmium orange is also shown by Methuen (1967) and Winsor & Newton. Orange is shown by Palmer (1962). It may seem to some that I am taking undue liberties by omitting both traditional names and calling them Spectrum Orange.

Orange is not acceptable as a name for the same reasons that make red, blue, green, etc., unacceptable. Cadmium is not desirable because it is not clearly descriptive. The name derives from the color of the pigment cadmium sulphide (CdS), but the sulphide varies widely in color. The name is often compounded with yellow, orange, lemon, and other colors, and I prefer to avoid it.

The extended tabulation shows about a dozen different colors. Many of the swatches were measured by spectrophotometer. While both visual and notational distinctions can be made throughout the list, it seems best to consider them as all belonging to the one "family" of Spectrum Orange.

Ridgway uses Cadmium Orange and Orange frequently in Bulletin 50, and he undoubtedly had distinctive colors in mind, but the citations indicate material confusion. He also frequently relates the colors to orange yellow, which is presumably a descriptive term, as there is no color of that name in his color guides. He also sometimes relates the orange colors to his Cadmium Yellow, called Orange Yellow (Color 18) in the COLOR GUIDE. It is almost impossible to list the various citations under their Ridgway names; some of them are shown under more than one color name.

The COLOR GUIDE swatch chosen for Spectrum Orange has a 5.0YR 6.5/16.0. notation, which does not vary unacceptably from Ridgway's Cadmium Orange and Palmer's Orange.

Notations of Colors Related to Spectrum Orange

Color Name	Ridgway Notation	Hamly Notation	Villalobos Notation
Cadmium Orange	III 13	5.0 YR 6.0/14.0	OOS-14-12°
Cadmium Orange, measure of 1912		4.5 YR 6.5/13.5*	
Cadmium Orange, Winsor & Newton		4.0 YR 6.6/14.0*	
Mikado Orange	III 13b	5.0 YR 7.0/12.0	OOS-15-11°
Xanthine Orange	III 13i	5.0 YR 5.2/12.0	OOS-11-12°
Spectrum Orange, Guide swatch		5.0 YR 6.5/16.0*	
Orange	III 15	5.2 YR 6.7/14.0	O-15 to 17-12°
Orange, measure of 1912		6.7 YR 6.7/13.1*	
Orange, measure of 1886		6.7 YR 7.2/13.3*	
Orange, Palmer glossy		5.7 YR 6.7/15.4*	O-17-12°
Orange, Palmer matte		5.8 YR 6.8/13.2*	
Cadmium Yellow Deep, Winsor & Newton		5.9 YR 6.9/14.0*	
Indian Yellow, Winsor & Newton		5.7 YR 7.0/12.8*	

Citations for Cadmium Orange

Vol. V   p. 778: underparts pure orange (between cadmium orange and saturn red, but nearer former)
♂Orange-bellied Trogon, *Trogonurus a. aurantiiventris*

Vol. IV   p. 734: head (partly), throat, chest intense orange (cadmium orange)
♂Salvin's Manakin, *Manacus aurantiacus*

Vol. II   p. 575: chin, throat, chest rich orange or cadmium orange (in spring)
♂Blackburnian Warbler, *Dendroica blackburniae*

Vol. II   p. 299: head, neck, chest, etc., rich cadmium orange or orange-yellow
Orange Oriole, *Icterus auratus*

Vol. II   p. 291: head, neck, chest, etc., intense orange, usually cadmium orange
♂Fiery Oriole, *Icterus cucullatus igneus*

Vol. II   p. 287: head, neck, chest, etc., varying cadmium yellow to cadmium orange (most intense on head, neck, chest areas) (in summer)
♂Hooded Oriole, *Icterus c. cucullatus*

*Vol. II   p. 283: upper head, neck rich cadmium orange
Spotted-breasted Oriole, *Icterus p. pectoralis*

*Vol. II   p. 95: head, neck, underparts cadmium orange (pileum is more intense, near orange chrome)
♂Swainson's Tanager, *Piranga b. bidentata*

49

Citations for Orange

The orange yellow of these citations is not a Ridgway color, but is merely a descriptive term.

Vol. VII   p. 164: mainly side of head deep orange yellow (cadmium yellow to orange)
Curaçao Paroquet, *Eupsittula p. pertinax*

*Vol. VII   p. 122: underparts rich orange yellow (exactly the orange shown in *Nomenclature*)
Blue and Yellow Macaw, *Ara ararauna*

Vol. V   p. 787: underparts rich pure orange yellow (cadmium yellow to orange)
♂Gartered Trogon, *Chrysotrogon caligatus* (♀paler)

Vol. V   p. 778: underparts pure orange (between cadmium orange and saturn red, but nearer former)
♂Orange-bellied Trogon, *Trogonurus a. aurantiiventris* (♀paler)

Vol. IV   p. 748: head, neck intense orange yellow (cadmium yellow to orange of *Nomenclature*)
♂Yellow-headed Manakin, *Pipra. e. erythrocephala*

Vol. II   p. 575: chin, throat, chest rich orange or cadmium orange (in spring)
♂Blackburnian Warbler, *Dendroica blackburniae*

*Vol. II   p. 314: part of head, neck orange or orange yellow (varying, with average hue about orange of *Nomenclature*); underparts similar but paler orange hues (in summer)
♂Bullock's Oriole, *Icterus bullockii*

Vol. II   p. 311; rump, wing and tail coverts, underparts rich orange or orange yellow (varying cadmium yellow to intense orange or almost flame scarlet, with average as cadmium orange) (in summer)
♂Baltimore Oriole, *Icterus galbula*

*Vol. II   p. 283: below rich orange yellow or orange
Spotted-breasted Oriole, *Icterus p. pectoralis*

*Vol. I   p. 623: head, neck, below orange
♂Orange-colored Grosbeak, *Pheucticus aurantiacus*

Orange Yellow ~ Color 18

Orange Yellow replaces Ridgway's Cadmium Yellow as a color name, despite my reluctance to differ from Ridgway's usage. As mentioned under Spectrum Orange (Color 17), the word *cadmium* does not denote any specific color quality. Ridgway combines it with both orange and yellow and relates these combinations to orange, chrome and chrome yellow, and lemon yellow. Of course, such usage does not diminish their value as color names, and the various colors are no doubt present in the plumages described.

However, Cadmium Yellow can best be thought of as a color between orange and yellow, and Orange Yellow is certainly an adequate name. It so happens that Ridgway frequently uses orange yellow as a descriptive term in Bulletin 50, although not as a specific color name.

Palmer (1962) depicts a color called Orange Yellow. It has a quality close to Ridgway's Cadmium Yellow and to the COLOR GUIDE swatch for Orange Yellow, which has a 10.0YR 8.0/14.0 notation.

Notations of Colors Related to Orange Yellow

Color Name	Ridgway Notation	Hamly Notation	Villalobos Notation
Cadmium Yellow	III 17	0.5 Y 7.2/13.0	OOY-16-12°
Cadmium Yellow, measure of 1912		9.5 YR 6.9/12.2*	
Cadmium Yellow, measure of 1886		10.1 YR 7.6/12.5*	
Deep Chrome	III 17b	9.5 YR 6.8/12.0	OOY-17-12°
Deep Chrome, measure of 1912		9.0 YR 7.2/10.2*	
Cadmium Yellow, Winsor & Newton		10.0 YR 7.8/14.4*	
New Gamboge, Winsor & Newton		0.5 Y 7.7/13.1*	
Orange Yellow, Palmer glossy		1.5 Y 7.5/14.0*	
Orange Yellow, Palmer matte		1.0 Y 7.5/15.0*	
Orange Yellow, Guide swatch		10.0 YR 8.0/14.0*	

Citations for Cadmium Yellow

*Vol. VII   p. 231: hindneck light cadmium yellow
Yellow-naped Parrot, *Amazona auro palliata*

Vol. VII   p. 164: side of head mainly cadmium yellow or orange (deep orange yellow)
Curaçao Paroquet, *Eupsittula p. pertinax*

Vol. VI   p. 119: forehead light chrome to cadmium yellow
♂Pucheran's Woodpecker, *Tripsurus p. pucherani*

*Vol. VI   p. 81: postnasal and forehead cadmium yellow
♂Golden-fronted Woodpecker, *Centurus aurifrons*

*Vol. VI	p. 78: prefrontal area cadmium yellow (more or less deep)
	♂Oaxaca Woodpecker, *Centurus p. polygrammus*
*Vol. VI	p. 15: shafts of primaries and rectrices bright cadmium yellow
	Flicker, *Colaptes a. auratus*
Vol. V	p. 787: below cadmium yellow to orange (rich pure orange yellow)
	♂Gartered Trogon, *Chrysotrogon caligatus*
Vol. V	p. 781: lower breast orange-yellow (between cadmium and chrome yellow)
	♂Graceful Trogon, *Trogonurus curucuitenellus*
Vol. V	p. 756: lower breast rich orange-yellow (cadmium or deep chrome)
	♂Black-headed Trogon, *Trogon m. melanocephalus*
Vol. IV	p. 748: head and neck cadmium yellow to orange (intense orange-yellow)
	♂Yellow-headed Manakin, *Pipra e. erythrocephala*
Vol. IV	p. 732: neck, throat, chest bright orange-yellow (between chrome and cadmium)
	♂Gould's Manakin, *Manacus vitellinus*
Vol. II	p. 311: rump, underparts varying from cadmium yellow to intense orange
	♂Baltimore Oriole, *Icterus galbula*
Vol. II	p. 305: pileum, nape more saffron or cadmium yellow
	Yellow-tailed Oriole, *Icterus m. mesomelas*
Vol. II	p. 301: head, neck, below brightest rich lemon or cadmium yellow
	Yellow Oriole, *Icterus x. xanthornus*
*Vol. II	p. 284: neck, rump, below rich cadmium yellow
	Lichtenstein's Oriole, *Icterus g. gularis*

Citations for Orange Yellow

Orange Yellow is not a Ridgway color, but the following are further citations of his use of the term.

Vol. VII	p. 164: side of head deep orange yellow (cadmium yellow to orange)
	Curaçao Paroquet, *Eupsittula p. pertinax*
Vol. VII	p. 122: below rich orange yellow (exactly orange shown in *Nomenclature*)
	Blue and Yellow Macaw, *Ara ararauna*
Vol. V	p. 787: below rich pure orange yellow (cadmium yellow to orange) ♀paler
	♂ Gartered Trogon, *Chrysotrogan caligatus*
Vol. V	p. 781: lower breast orange yellow (between cadmium and chrome yellow)
	♂Graceful Trogon, *Trogonurus curucuitenellus*
Vol. V	p. 756: lower breast rich orange yellow (cadmium or deep chrome)
	♂Black-headed Trogon, *Trogon m. melanocephalus*

Dusky Brown ~ Color 19
Dark Grayish Brown ~ Color 20

Dusky Brown and Dark Grayish Brown are shown in *Color Standards*, where the swatches are very similar. They approach Fuscous (Color 21), but they have a slightly redder hue, whereas Fuscous is visibly more yellowish orange. All are dark and display little chromaticity. In this area of low intensity, the hues are difficult to see, and consequently many such colors can be related to each other.

Colors related to Dusky Brown are seldom used in Bulletin 50, and no useful citations were found for them, with the exception of Seal Brown and Dark Grayish Brown. Dark Grayish Brown is used with great frequency in Bulletin 50, but usually in a generalized descriptive sense; only two specific citations were found.

The glossy finish of Palmer's (1962) swatch for Dusky Brown measured 7.5R 3.0/1.5 on the spectrophotometer, and the matte swatch measured 7.0R 3.6/1.0. Palmer does not use the Munsell system to identify his swatches, but instead uses the Villalobos system. In his pamphlet (Palmer and Reilly, 1956) listing the Villalobos notations defining his colors, Palmer's notation for Dusky Brown is R-6-2°, which is the same notation Villalobos gives for Ridgway's Dusky Brown. The Palmer and Ridgway swatches should match, but they do not; Palmer's is clearly redder.

Palmer also shows a swatch called Blackish Brown, which measured a 7.5R 2.5/1.4 notation, close to that found for his Dusky Brown. Villalobos gives it a notation of S-4-2°, which is close to his notation for Ridgway's Light Seal Brown.

A study of two birds cited in Bulletin 50 for Dusky and Dark Grayish Brown resulted in two colors to represent this color range. They are approximately midway between the colors indicated by the Ridgway and Palmer notations.

The COLOR GUIDE swatch chosen for Dusky Brown has a 4.0R 2.5/0.7 notation; the swatch for Dark Grayish Brown has a 6.0R 2.5/1.0 notation.

Notations of Colors Related to Dusky Brown

Color Name	Ridgway Notation	Hamly Notation		Villalobos Notation
Dusky Brown	XLV 1''''k	8.0 RP	3.2/1.5	R-6-2°
Dusky Brown, measure of 1912		1.3 R	3.1/1.2*	
Dusky Brown, Palmer glossy		7.5 R	3.0/1.5*	R-6-2°
Dusky Brown, Palmer matte		7.0 R	3.6/1.0*	R-6-2°
Dusky Brown, Guide swatch		4.0 R	2.5/0.7*	
Blackish Brown #1	XLV 1''''m	10.0 RP	1.0/3.0	MR-4-2°
Blackish Brown, Palmer		7.5 R	2.5/1.4*	S-4-2°
Warm Blackish Brown	XXXIX 1'''m	2.5 R	3.4/1.5	R-5-2°
Dark Livid Brown	XXXIX 1'''k	2.5 R	3.6/2.0	R-6-3°
Dark Vinaceous-Drab	XLV 5''''i	2.5 R	3.8/1.5	S-7-3°
Dark Grayish Brown	XLV 5''''k	2.5 R	3.8/1.5	S-6-2°
Dark Grayish Brown, Guide swatch		6.0 R	2.5/1.0*	
Blackish Brown #2	XLV 5''''m	2.5 R	2.8/1.0	S-4-2°
Blackish Brown #2, measure of 1912		1.2 YR	2.9/0.5*	
Seal Brown	XXXIX 5'''m	7.5 R	3.2/1.5	RS-5-2°
Light Seal Brown	XXXIX 9'''m	10.0 R	2.6/1.0	S-4-3°

Citations for Seal Brown

Vol. V   p. 119: upper tail coverts seal brown or dark chocolate
Black-headed Antthrush, *Formicarius analis nigricapillus*

Vol. V   p. 114: above dark chocolate, or between that and seal brown
♀Zeledon's Antbird, *Myrmeciza zeledoni*

Vol. II   p. 152: above dark chocolate, varying to seal brown
Dusky-tailed Ant Tanager, *Phoenicothraupis fuscicauda*

Citations for Dusky and Dark Grayish Brown

Vol. VIII p. 353: primaries dusky grayish brown (blackish fuscous)
Green Sandpiper, *Tringa ochrophus*

Vol. IV   p. 163: wings, rectrices dark grayish brown or dusky
Townsend's Solitaire, *Myadestes townsendi*

Fuscous ~ Color 21

Fuscous is defined in a dictionary as any of several colors that average red yellow in hue with low saturation and brilliance. When compared with Dusky Brown, Fuscous has less of the reddish hue and emphasizes yellow. Both colors are so dark that the hues are difficult to discern.

Fuscous is shown in *Color Standards,* but not in *Nomenclature,* which may indicate that the name was a later conception of Ridgway's. The inference is somewhat confirmed by his writings, where Fuscous is used only twice in the first five volumes of Bulletin 50, although there are frequent opportunities to use it, and it is used increasingly in the later volumes.

Ridgway uses three generalized descriptive terms in Bulletin 50 related to Fuscous: dusky, grayish brown, and sooty brown. None of these is identified by a color swatch in *Color Standards.*

Dusky is a term used almost in the sense of a specific color rather than merely the equivalent of dark or deep. The wings and tails of birds are areas most frequently so described. Dusky is sometimes equated with dull black, dull blackish gray, or dark grayish brown. Twice, however, Ridgway speaks of it as "nearly clove brown," which is included in the Fuscous range. Dusky may be thought of as a quality between Fuscous and Dusky Brown, but examination of plumages called dusky has led me to consider it almost synonymous with Fuscous. Only a few of the many citations are listed.

Grayish brown is used even more often than dusky, but the only swatch shown in *Color Standards* is Dark Grayish Brown (Color 20). Grayish brown is given a wide latitude, from very dark tones to others as light as Isabella Color. Therefore, Ridgway usually attempts to clarify his meaning by the addition of modifying words such as *deep, dark,* and *light.* As a result, the term *grayish brown* has little, if any, value.

Sooty brown seems to be thought of as a color between Ridgway's Fuscous Black and Clove Brown at its darkest extreme, or between Clove Brown and Dark Sepia at its palest—both are rather dark. There are a few examples in the citations for Clove Brown, and others are shown for Sepia (see Olive-Brown, Color 28).

My interest in Fuscous as an effective color name was greatly influenced by its recent use by Wesley Lanyon and others (Lanyon, 1960, 1961; Phillips, 1966, 1970) in their descriptions of the tails of certain flycatchers of the genus *Myiarchus*. The authors use the patterns of the dusky areas of the rectrices (which they call fuscous) and the adjacent pale brown areas (which they call cinnamon-rufous) to aid in the identification of species and subspecies. Ridgway does not call the dark areas of these flycatchers fuscous, but speaks of them as deep brownish olive and deep or dark grayish brown. I made very detailed studies of these areas and of color swatches and came to the conclusion that the designation of the dark areas as fuscous is an excellent choice.

My interest in the *Myiarchus* papers was also influenced by a study of Palmer's glossy swatch for Fuscous. It is different from that shown in the NATURALIST'S COLOR GUIDE and from the *Myiarchus* specimens: substantially paler with a strong trend toward Ridgway's Olive-Brown. Palmer's matte swatch registers somewhat closer, but it is still too pale and intense.

The tabulation includes seven other Ridgway colors, all of substantial grayishness and low chromaticity. There is a visible difference between Blackish Brown #3 and Clove Brown, but their divergence from Fuscous is not material to this effort to eliminate an excessive quantity of color names. Fuscous, plus modifying words such as *darker, paler, grayer,* or *yellower,* should suffice in this dark color area.

The COLOR GUIDE swatch chosen for Fuscous has a 5.0YR 3.0/1.3 notation.

Notations of Colors Related to Fuscous

Color Name	Ridgway Notation	Hamly Notation	Villalobos Notation
Fuscous	XLVI 13''''k	5.0 YR 3.5/1.0	SO-6-2°
Fuscous, measure of 1912		7.5 YR 3.4/1.1*	
Fuscous, Guide swatch		5.0 YR 3.0/1.3*	
Fuscous, Palmer glossy		7.5 YR 3.7/2.4*	SO-6-2°
Fuscous, Palmer matte		6.5 YR 4.1/1.9*	SO-6-2°
Dusky Drab	XLV 9''''k	5.0 YR 3.5/1.0	SSO-6-2°
Blackish Brown #3	XLV 9''''m	5.0 YR 2.5/1.0	S-4-2°
Blackish Brown #3, measure of 1912		2.1 YR 2.9/0.5*	
Fuscous Black	XLVI 13''''m	5.0 YR 2.6/0.5	S-4-2°
Bone Brown	XL 13'''m	5.0 YR 2.6/1.5	S-5-2°
Clove Brown	XL 17'''m	10.0 YR 2.8/1.5	SSO-4-2°
Chaetura Drab	XLVI 17''''k	10.0 YR 3.6/1.0	OOS-6-3°
Chaetura Drab, measure of 1912		1.8 YR 3.4/1.0*	
Chaetura Black	XLVI 17''''m	10.0 YR 2.4/1.0	SSO-4-1°

Citations for Fuscous

Vol. XI — p. 483: back, breast, etc., varying from fuscous to chaetura drab, even to fuscous black
Bald Eagle, *Heliaeetus l. leucocephalus*

Vol. XI — p. 280: upper parts fuscous to fuscous black
Red-shouldered Hawk, *Buteo l. lineatus*

Vol. XI — p. 246: head and body fuscous to chaetura drab (melanistic)
Western Red-tailed Hawk, *Buteo jamaicensis calurus*

*Vol. XI — p. 132: back fuscous plus other details
Snail Kite, *Rostrhamus s. sociabilis*

Vol. XI — p. 104: crown, occiput fuscous to fuscous black (♀ in brown phase, ♂ melanistic, deep fuscous black)
Hookbilled Kite, *Chondrohierax u. uncinatus*

Vol. VIII — p. 721: back grayer than fuscous, nearest chaetura drab
Murre, *Uria t. trolle*

Vol. VIII — p. 397: pileum fuscous or clove brown
Whimbrel, *Phaeopus p. phaeopus*

Vol. VII — p. 292: back, scapulars fuscous or chaetura drab
Costa Rican Bandtailed Pigeon, *Chloroenas albilinea crissalis*

*Vol. V — p. 233: margins throat and midbreast fuscous
Spot-necked Woodhewer, *Dendrocolaptes puncticollis*

*Vol. II — p. 473: wings and tail fuscous
♀ Colina Warbler, *Helminthophila crisallis*

Citations for Dusky Drab

*Vol. VIII — p. 443: occiput dusky drab, browner than slate gray crown
Grand Cayman Dove, *Leptotila collaris*

Citations for Blackish Brown #3 and Chaetura Black

No useful citations for Blackish Brown #3 or Chaetura Black were found in Bulletin 50.

Citations for Fuscous Black

See citations for Fuscous (Vol. XI, pp. 483, 280, 103) and Clove Brown (Vol. VIII, p. 55).

Vol. VIII — p. 32: primaries fuscous black to chaetura drab
American Oystercatcher, *Haematopus p. palliatus*

Citations for Bone Brown

Vol. VI — p. 624: upper parts bister to nearly bone brown
Richardson's Owl, *Cryptoglaux tengmalmi richardsoni*

Citations for Chaetura Drab

See citations for Fuscous (Vol. XI, pp. 483, 246; Vol. VIII, p. 721; Vol. VII, p. 292) and Fuscous Black (Vol. VIII, p. 32).

Vol. VIII — p. 353: above deep grayish brown, nearest chaetura drab
Green Sandpiper, *Tringa ochrophus*

Citations for Clove Brown

*Vol. VIII   p. 707: head, neck, chest dark sooty brown, clove brown
Dovekie, *Plautus alle*

Vol. VIII   p. 55: upper parts dark sooty clove brown to fuscous black
Black Turnstone, *Arenaria melanocephala*

*Vol. VIII   p. 40: above dark sooty brown, nearest clove brown
Black Oystercatcher, *Haematopus bachmani*

*Vol. V   p. 324: pileum dusky, nearly clove brown
Island Hermit, *Phaethornis anthophilus hyialinus*

*Vol. V   p. 319: pileum dusky, nearly clove brown
Guatemalan Hermit, *Phaethornis l. longirostris*

Vol. IV   p. 870: part of head grayish warm sepia to between clove brown and seal brown
♀Costa Rican Tityra, *Tityra semifasciata costaricensis*

Vol. IV   p. 711: pileum sooty brown, between sepia and clove brown
Brown-capped Tyrannulet, *Microtriccus brunneicapillus*

Vol. IV   p. 665: pileum dark sooty brown, clove brown
Boat-billed Flycatcher, *Megarynchus pitangua mexicanus*

Vol. IV   p. 269: above deep sooty brown, between sepia and clove brown; wings and tail darker blackish brown or clove brown
White-breasted Trembler, *Ramphocinclus brachyurus*

Vol. IV   p. 136: dark sooty brown, between sepia and clove brown
♂Aztec Thrush, *Ridgwayia pinicola*

Citations for Sooty and Sooty Brown

See Clove Brown citations (Vol. VIII, pp. 707, 55, 40; Vol. IV, pp. 711, 665, 269, 136).

Citations for Dusky

A limited selection of citations is given; many more are available. See Clove Brown citations (Vol. V, pp. 324, 319).

*Vol. IV   p. 561: wings dusky
Least Flycatcher, *Empidonax minimus*

Vol. IV   p. 163: wings dark grayish brown or dusky
Townsend's Solitaire, *Myadestes townsendi*

*Vol. III   p. 595: wings and tail dusky
♂Blackpoll Warbler, *Dendroica striata*

Vol. III   p. 592: tail grayish black or dusky
♂Bay-breasted Warbler, *Dendroica castanea*

Vol. I   p. 80: wings and tail grayish dusky
♀Greenland Redpoll, *Acanthus h. hornemanni*

*Vol. I   p. 68: wings and tail dusky
Gray-crowned Rosy Finch, *Leucosticte t. tephrocotis*

Burnt Umber ~ Color 22

Burnt Umber is one of twelve Ridgway colors shown in *Color Standards* and gathered into one "family" in the NATURALIST'S COLOR GUIDE. Burnt Umber, Chocolate, Vandyke Brown (also shown in *Nomenclature*), and Chestnut-Brown are the names most often used in Bulletin 50. In no instance, however, are they used without being related to other colors. Despite the large number of names available to him in this color range, Ridgway is unable to choose one unequivocal color to describe a plumage. Sometimes this may be due to the inherent variations in the plumages described, but often it occurs because a match is impossible, confirming my belief in the necessity of reducing the number of colors shown in any color guide.

The choice of a color name from among the twelve of this group presented a problem. Only three were worth considering. Chocolate embraces too wide a range; milk chocolates differ too greatly from dark, bitter chocolates. Chestnut-Brown is too readily confused with Chestnut, unless invariably hyphenated. Burnt Umber, on the other hand, is a commercially known paint color with a limited, dark brownish tonal range. It seems to be the best choice.

The tabulation is divided into two sections in order to separate the YR hues from the slightly ruddier R hues. However, the notations given by Villalobos indicate a very close relationship.

The COLOR GUIDE swatch chosen for Burnt Umber has a 2.5YR 3.0/2.5 notation.

Citations for Burnt Umber

Vol. VI	p. 647: below general color dark burnt umber or dark vandyke brown
	Guatemalan Barred Owl, *Strix fulvescens*
Vol. V	p. 132: above, including pileum and nape, chestnut-brown or between prout's brown and vandyke, even nearly burnt umber
	Olive-sided Antbird, *Anoplops olivascens*
Vol. V	p. 107: crown and nape burnt umber or vandyke
	♂ White-bellied Antbird, *Myrmeciza boucardi panamensis*
Vol. III	p. 531: upper parts chestnut-brown, or between burnt umber and mummy brown
	Panama Black-bellied Wren, *Pheugopedius fasciato-ventris albigularis*
Vol. II	p. 243: above between burnt umber and prout's brown
	♀ Slender-billed Grackle, *Megaquiscalus tenuirostris*

Notations of Colors Related to Burnt Umber

Color Name	Ridgway Notation	Hamly Notation	Villalobos Notation
Burnt Umber	XXVIII 9″m	2.5 YR 3.0/2.0	SSO-5-3°
Burnt Umber, measure of 1912		2.0 YR 3.2/2.4*	
Burnt Umber, measure of 1886		4.0 YR 3.9/4.8*	
Burnt Umber, Guide swatch		2.5 YR 3.0/2.5*	
Chestnut-Brown	XIV 11′m	2.5 YR 3.4/3.0	SO-5-3°
Chestnut-Brown, measure of 1912		4.0 YR 3.3/3.0*	
Chocolate	XXVIII 7″m	2.5 YR 2.6/2.5	S-5-4°
Vandyke Brown	XXVIII 11″m	5.0 YR 3.5/2.5	SSO-5-3°
Warm Sepia	XXIX 13″m	5.0 YR 3.8/2.0	SO-6-3°
Warm Sepia, measure of 1912		4.9 YR 3.4/2.1*	
Carob Brown	XIV 9′m	2.5 YR 2.0/4.0	SSO-5-4°

Supplementary Table

Dark Vinaceous-Brown	XXXIX 5‴k	7.5 R 3.8/2.0	S-6-4°
Diamine Brown	XIII 3′m	7.5 R 3.2/2.5	S-5-4°
Hessian Brown	XIII 5′m	10.0 R 2.5/2.5	S-5-5°
Sorghum Brown	XXXIX 9‴i	10.0 R 3.0/2.0	SO-9-3°
Hay's Brown	XXXIX 9‴k	10.0 R 4.5/2.0	SSO-5-3°
Liver Brown	XIV 7′m	10.0 R 3.0/3.5	SSO-5-4°

Citations for Chestnut-Brown

Vol. V	p. 132: above chestnut-brown or vandyke brown
	Bicolored Antbird, *Anoplops bicolor*
Vol. V	p. 132: above, including pileum and nape, chestnut-brown, or between prout's brown to vandyke, even nearly burnt umber
	Olive-sided Antbird, *Anoplops olivascens*
Vol. V	p. 113: flanks, under tail chestnut-brown or vandyke brown
	♂Cherrie's Antbird, *Myrmeciza exsul occidentalis*
*Vol. V	p. 107: crown and nape burnt umber or vandyke, sometimes uniform chestnut-brown
	♂White-bellied Antbird, *Myrmeciza boucardi panamensis*
Vol. V	p. 56: pileum chestnut-brown or russet
	♀Northern Antvireo, *Dysithamnus mentalis septentrionalis*
*Vol. III	p. 532: above richer chestnut-brown, back decidedly chestnut
	Black-bellied Wren, *Pheugopedius fasciato-ventris melanogaster*
Vol. III	p. 531: upper parts chestnut-brown, or between burnt umber and mummy
	Panama Black-bellied Wren, *Pheugopedius fasciato-ventris albigularis*

Citations for Chocolate

Vol. IX   p. 169: lower back, rump chocolate brown to clove brown
                     Guatemalan Ruddy Rail, *Latteralis r. ruber*

*Vol. IX   p. 168: most of back, wing-coverts dark chocolate brown
                     Tamaulipas Ruddy Rail, *Latteralis ruber tamaulipensus*

Vol. V   p. 119: upper tail-coverts seal brown or dark chocolate
                     Black-headed Antthrush, *Formicarius analis nigricapillus*

Vol. V   p. 114: above dark chocolate or between that and seal brown;
                     underparts lighter chocolate or vandyke brown
                     ♀Zeledon's Antbird, *Myrmeciza zeledoni*

Vol. II   p. 152: above dark chocolate, varying to seal brown
                     ♂Dusky-tailed Ant Tanager, *Phoenicothraupis fuscicauda*

Citations for Vandyke Brown

See citations for Burnt Umber (Vol. V, p. 132; Vol. VI, p. 647) and Chestnut-Brown (Vol. V, pp. 107, 113, 132).

Vol. V   pp. 118, 119: above general color dark vandyke brown or
                     approaching seal brown
                     Black-headed Antthrush, *Formicarius analis nigricapillus*

Vol. V   p. 111: underparts vandyke brown or mummy brown
                     ♀Sclater's Antbird, *Myrmeciza e. exsul*

Vol. IV   p. 849: pileum and nape deep brown, prout's brown to between
                     vandyke brown and sepia
                     ♀Jamaican Becard, *Platypsaris niger*

Citations for Warm Sepia

Vol. V   p. 217: above warm sepia or deep olive brown
                     Buff-throated Foliage Gleaner, *Automolus cervinigularis*

Raw Umber ~ Color 23

The Raw Umber group of colors is composed of Ridgway's Raw Umber, Mars Brown, Prout's Brown, and Mummy Brown. Raw Umber was chosen to represent the lot because it is a commercially available paint with a comparatively narrow tonal range, and its color name is therefore more meaningful than any of the others. Variations from the basic color swatch can readily be noted by the use of modifying terms such as *lighter, darker, ruddier,* etc.

The group is related to Olive-Brown (Color 28), from which it has been separated because of a clearly more intense chroma. The separation is also confirmed by the relative positions of the colors in Ridgway's *Color Standards*. The relationship between the two groups is indicated by a reading of the citations for Raw Umber, many of which also mention Olive-Brown. The Villalobos notations make very little distinction between Raw Umber, Burnt Umber, and Olive-Brown and come close to equating some of the colors of all three groups. On the other hand, Hamly's notations consistently separate the colors into three groups. This may be another example of the different results obtained by different authors when they measure swatches in different copies of *Color Standards*.

Horn Color deserves special discussion. It occurs with great frequency in Bulletin 50, and it is commonly used by ornithologists describing the color of bills, legs, toes, and nails of birds. It is a moderately dark brown color with a slightly grayish tone related to Raw Umber.

Ridgway does not show a swatch for Horn Color. He mentions it in *Nomenclature,* but only to correlate the name in different languages: horn color in English, *hornfarbig* in German, *color de cuerno* in Spanish, etc. He does not define it, and it seems never to have been defined. Nevertheless, everyone who uses it apparently believes he understands the color.

In this discussion, horns are those of domesticated cattle. Horns are made of tough, fibrous material consisting chiefly of keratin, the chemical basis of epidermal tissues and exteriors of horns, hair, nails, and even feathers. The word *horn* in Horn Color indicates that some sort of texture is an important element, perhaps more important than the precise color. The texture most obviously common to horns and horn-colored parts of birds is a distinctively polished surface, a shine or sheen peculiar to horns. It is possible that some characteristic color (possibly grayish white) was at one time associated with horns, but there is no clear evidence of this available.

A color, or some range of colors, may be inferred from Ridgway's descriptions, of which hundreds are available in Bulletin 50. A few typical selections follow:

Bill horn color, legs similar but paler
Bill horn color, tip more dusky
Bill horn color, or grayish brown
Bill horn color, or browner, in winter
Bill horn color, or grayer, in life
Bill blackish, but paler, more horn-colored basally
Bill dusky horn color or blackish brown
Legs deep horn color or blackish brown
Mandible dusky, with stripe of horn color along tomium
Maxilla blackish brown or deep horn color
Grayish-horn, yellowish-horn, bluish-horn, etc.

From these and other descriptions, I derive the impression that the color is basically brownish, moderately dark but not blackish, and possibly with some grayish tone. Horn Color—apart from its characteristic texture—seems to approach Raw Umber most closely.

The COLOR GUIDE swatch chosen for Raw Umber has a 7.5YR 3.5/3.0 notation.

Notations of Colors Related to Raw Umber

Color Name	Ridgway Notation	Hamly Notation	Villalobos Notation
Mars Brown	XV 13'm	5.0 YR 3.4/3.5	SO-6-3°
Raw Umber	III 17'm	7.5 YR 3.5/3.0	O-5-5°
Raw Umber, measure of 1912		10.0 YR 3.4/3.2*	
Raw Umber, Guide swatch		7.5 YR 3.5/3.0*	
Prout's Brown	XV 15'm	7.5 YR 3.5/3.0	OOS-5-5°
Prout's Brown, measure of 1912		6.5 YR 3.5/2.8*	
Prout's Brown, measure of 1886		6.8 YR 4.2/3.8*	
Mummy Brown	XV 17'm	10.0 YR 3.4/3.0	SO-4-2°

Citations for Mars Brown

See citations for Raw Umber (Vol. IV, p. 271); Mummy Brown (Vol. V, p. 110); and Raw Umber (Vol. IV, pp. 28, 32).

Vol. V   p. 229: above olive brown, nearly raw umber to mars brown
           Barred Woodhewer, *Dendrocolaptes s. sancti-thomae*
Vol. IV   p. 26: pileum, nape deep russet, or between mars brown and russet
           Russet Nightingale Thrush, *Catharus o. occidentalis*

Citations for Raw Umber

A substantial number of citations for Raw Umber also refer to Olive-Brown, often with some emphasis on the latter. It was thought best to detail them under Olive-Brown (Vol. II, p. 125; Vol. III, p. 534, Vol. IV, pp. 28, 52, 115, 591; Vol. V, pp. 85, 121, 135, 185, 229, 233, 250, 254, 285).

*Vol. V   p. 781: above brown, nearly raw umber, pileum deeper, rump paler
       ♀Graceful Trogon, *Trogonurus curucuitenellus*

Vol. V   p. 779: above nearest raw umber, or between that and mummy brown
       ♀Orange-billed Trogon, *Trogonurus a. aurantiventris*

Vol. V   p. 776: pileum brown, between raw umber and mummy, to nearly russet
       ♀Jalapa Trogon, *Trogonurus puella*

Vol. V   p. 768: above brown, between raw umber and isabella color
       ♀Coppery-tailed Trogon, *Trogonurus a. ambiguus*

Vol. V   p. 291: above between raw umber and mummy, to nearly russet
       Brown Dendrocincla, *Dendrocincla lafresnayei ridgwayi*

Vol. V   p. 288: pileum, nape sepia to near raw umber, back and scapulars sometimes raw umber, underparts light raw umber
       Northern Dendrocincla, *Dendrocincla a. anabatina*

Vol. V   p. 272: above deep brown, between raw umber and mummy brown
       Costa Rican Sickle Bill, *Campyloramphus borealis*

Vol. V   p. 271: pileum, nape light sepia or dark raw umber; back more fulvous or russet, between raw umber and russet or mars brown
       Venezuelan Sickle Bill, *Campylorhamphus venezuelensis*

*Vol. V   p. 261: above raw umber, but with narrow streaks
       Allied Woodhewer, *Picolaptes a. affinis*

Vol. V   p. 209: above between raw umber and mummy, but with streaks
       Lineated Xenicopsis, *Xenicopsis subalaris lineatus*

Vol. V   p. 237: above near raw umber or between it and light olive
       Guatemalan Woodhewer, *Xiphocolaptes e. emigrans*

*Vol. V   p. 616: above brown, raw umber
       Guerrero Automolus, *Automolus guerrerensis*

Vol. V   p. 207: above brown, between raw umber and mummy brown
       Scaly-throated Xenicopsis, *Xenicopsis variegaticeps*

*Vol. V   p. 203: nape, back tawny, nearly raw umber
       Ochraceous Philydor, *Philydor panerythrus*

Vol. V   p. 175: nape, back dull cinnamon, between russet and raw umber
       Streaked Xenops, *Xenops rutilus heterurus*

Vol. V   p. 22: above tawny, nearest raw umber or tawny olive
       ♂Tawny Antshrike, *Thamnistes a. anabatinus*

Vol. IV   p. 811: nape, upper back buffy brown, between wood brown and raw umber; scapulars, lower back deepening to raw umber
       ♂Gaumer's Attila, *Attila citreopygus gaumeri*

Vol. IV   p. 32: above dark raw umber to mummy or near mars brown
       Slender-billed Nightingale Thrush, *Catharus g. gracilirostris*

Vol. IV	p. 28: pileum, nape deep russet, near mars brown to raw umber; rest of upper parts olive brown or raw umber
	Frantzius' Nightingale Thrush, *Catharus f. frantzii*
Vol. IV	p. 26: above olivaceous russet to nearly raw umber
	Russet Nightingale Thrush, *Catharus o. occidentalis*
*Vol. I	p. 601: above nearly raw umber
	♀Blue Bunting, *Cyanocompsa p. parellina*

Citations for Prout's Brown

*Vol. VI	p. 401: pileum nearly prout's brown
	Panama Nonnula, *Nonnula frontalis*
Vol. V	p. 264: pileum sepia or light sepia, passing to lighter warm brown, or nearly prout's brown on hindneck
	Streak-headed Woodhewer, *Picolaptes l. lineaticeps*
Vol. V	p. 204: pileum deep chestnut, between prout's brown and mummy
	Dusky-winged Philydor, *Philydor fuscipennis*
Vol. V	p. 128: pileum, nape prout's brown to deep broccoli brown
	♀Spotted Antbird, *Hylophylax naevioides*
Vol. V	p. 85: pileum, nape olive brown, between prout's and raw umber
	Long-billed Antwren, *Ramphocaenus r. rufiventris* (Northern)
Vol. IV	p. 287: above deep broccoli to prout's brown or vandyke
	♂Scarlet-cheeked Weaver-Finch, *Estrilda melpoda*
Vol. III	p. 597: above rufescent brown, between prout's brown and mummy
	Winter Wren, *Olbiorchilus h. hiemalis*
Vol. III	p. 593: above between prout's brown and bister or sepia
	Martinique Wren, *Troglodytes martinicensis*
*Vol. III	p. 579: above nearest prout's brown in summer; deeper, nearly chestnut-brown in winter
	House Wren, *Troglodytes a. aedon*
Vol. III	p. 541: above rusty brown, near prout's brown to chestnut-brown in summer; almost chestnut in winter
	Carolina Wren, *Thryothorus l. ludoviciana*
Vol. II	p. 243: pileum, nape, back between prout's brown and burnt umber
	♀Slender-billed Grackle, *Megaquiscalus tenuirostris*
*Vol. I	p. 423: above dull prout's brown, below more raw umber
	♀Towhee, *Pipilo e. erythrophthalmus*

Citations for Mummy Brown

See citations for Raw Umber (Vol. IV, p. 32; Vol. V, pp. 207, 209, 272, 288, 779, 781).

Vol. X	p. 23: back, rump dull dusky sepia to mummy brown
	Northern Crested Guan, *Penelope p. purpurascens*
Vol. VI	p. 506: above general color between bister and mummy brown
	♂Chuck-will's-widow, *Antrostomus carolinensis*

Vol. V	p. 204: pileum deep chestnut-brown, between mummy and prout's
	Dusky-winged Philydor, *Philydor fuscipennis*
*Vol. V	p. 181: back mummy brown
	Costa Rican Premnoplex, *Premnoplex brunnescens brunneicauda*
*Vol. V	p. 119: back mummy brown
	Mexican Antthrush, *Formicarius m. moniliger*
Vol. V	p. 110: above bright mummy or mars brown
	♂Cassin's Antbird, *Myrmeciza cassini*
*Vol. V	p. 102: pileum, nape nearly mummy brown
	♀Bare-fronted Antbird, *Gymnocichla chiroleuca*
Vol. III	p. 597: above rufescent brown, between mummy and prout's brown
	Winter Wren, *Olbiorchilus h. hiemalis*
Vol. III	p. 590: above tawny, between mummy brown and russet
	Irazú Wren, *Troglodytes ochraceous*
Vol. III	p. 531: above chestnut-brown, between burnt umber and mummy
	♂Panama Black-bellied Wren, *Pheugopedius fasciato-ventris albigularis*
Vol. II	p. 124: above rich brown, between mummy brown and tawny olive
	♀White-throated Shrike-Tanager, *Lanio leucothorax*

Buff ~ Color 24

Buff is not shown by Ridgway, although he uses the term in more than twenty-four compound color names, such as Ochraceous-Buff, Buffy Brown, Buff-Yellow, etc. Buffy Brown is shown by Ridgway in *Color Standards* and by Palmer (1962). It is too dark to represent buffiness, and the NATURALIST'S COLOR GUIDE chooses a hue close to Ridgway's Ochraceous-Buff for that purpose. It is also close to the color of the nape of the Bobolink, the only Ridgway citation that unequivocally calls a plumage buff.

The COLOR GUIDE swatch chosen for Buff has a 1.0Y 6.0/7.0 notation.

Notations of Colors Related to Buff

Color Name	Ridgway Notation	Hamly Notation	Villalobos Notation
Ochraceous-Buff	XV 15'b	1.0 YR 8.0/6.0[1]	O-16-9°
Ochraceous-Buff, measure of 1912		9.6 YR 6.7/6.6*	
Ochraceous-Buff, measure of 1886		8.8 YR 7.6/6.2*	
Buff, Guide swatch		1.0 Y 6.0/7.0*	
Buffy Brown	XL 17'''i	10.0 YR 5.5/4.0	O-9-3°
Buffy Brown, measure of 1912		10.0 YR 4.7/3.0*	
Buffy Brown, Palmer		9.4 YR 4.5/4.4*	O-9-3°

[1] Hamly's 1.0YR may be a typographical error for 10.0YR.

Citations for Buff

Vol. VIII	p. 412: below dull buff or between pinkish-buff and vinaceous buff, in breeding plumage Eskimo Curlew, *Mesoscolopax borealis*
*Vol. VI	p. 385: nuchal collar deep buff, belly buff Fulvous Puffbird, *Ecchannornis radiatus flavidus*
Vol. V	p. 250: below buff, nearly isabella color or wood brown Lawrence's Woodhewer, *Xiphorhynchus n. nanus*
Vol. V	p. 244: below light buffy brown, nearly isabella color Swainson's Woodhewer, *Xiphorhynchus f. flavigaster*
Vol. V	p. 217: lower face, throat deep buff or ochraceous buff Buff-throated Automolus, *Automolus c. cervinigularis*
*Vol. IV	p. 840: below buff or tawny-buff, chest more ochraceous Cinnamon Becard, *Pachyrhamphus cinnamomeus*

*Vol. IV	p. 187: underparts pale buff, approaching buffy white on chin, throat, and abdomen, but entirely buff in fresh autumnal and early winter plumage
	Brown Thrasher, *Toxostoma rufum*
*Vol. II	p. 370: hindneck buff
	♂Bobolink, *Dolichonyx oryzivorus*

Citations for Ochraceous-Buff

Vol. VII	p. 76: throat, chest light ochraceous-buff or clay color
	Roadrunner, *Geococcyx californianus*
*Vol. VII	p. 64: lower breast, belly, thighs ochraceous-buff
	Haitian Lizard Cuckoo, *Saurothera dominicensis*
Vol. VII	p. 21: below dull ochraceous-buff or cinnamon-buff
	Mangrove Cuckoo, *Coccyzus m. minor*
Vol. V	p. 217: lower face, throat deep buff or ochraceous-buff
	Buff-throated Automolus, *Automolus c. cervinigularis*
Vol. V	p. 38: below nearly ochraceous-buff (between that and clay color)
	♀Black-crested Antshrike, *Thamnophilus radiatus nigricristatus*

69

Fawn Color ~ Color 25

Fawn Color is grouped in the NATURALIST'S COLOR GUIDE with Wood Brown, partly because of their close proximity in *Color Standards* and partly because the notations given by Villalobos are practically synonymous. The citations do not clarify the relationship between the colors very well.

The COLOR GUIDE swatch chosen for Fawn Color has a 7.5YR 5.0/3.0 notation, about midway between Hamly's notations for Ridgway's Fawn Color and Wood Brown.

Notations of Colors Related to Fawn Color

Color Name	Ridgway Notation	Hamly Notation	Villalobos Notation
Fawn Color	XL 13'''	5.0 YR 4.6/3.0	OOS-10-4°
Fawn Color, measure of 1912		4.4 YR 5.2/3.8*	
Fawn Color, measure of 1886		6.7 YR 6.5/3.9*	
Fawn Color, Guide swatch		7.5 YR 5.0/3.0*	
Wood Brown	XL 17'''	9.0 YR 6.0/3.0	OOS-11-4°
Wood Brown, measure of 1912		7.6 YR 5.4/3.3*	

Citations for Fawn Color

*Vol. VII   p. 353: forehead, sides of head and neck fawn color
       ♂Yucatan Mourning Dove, *Zenaidura yucatanensis*
*Vol. VII   p. 351: parts of head bright fawn, approaching sayal brown
       ♂Tres Marias Mourning Dove, *Zenaidura macroura tresmariae*
Vol. VII   p. 341: forehead fawn to deep avellaneous; below general color vinaceous-fawn, but chest deeper, nearly fawn
       ♂Cuban Mourning Dove, *Zenaidura m. macroura*

Citations for Wood Brown

Vol. VII   p. 357: ♂above in general deep wood brown, or snuff to buffy brown, ♀foreneck and chest nearly wood brown
       Zenaida Dove, *Zenaida z. zenaida*
Vol. V   p. 250: below wood brown to nearly isabella (equals light buffy brown)
       Lawrence's Woodhewer, *Xiphorhynchus n. nanus*
Vol. V   p. 58: pileum wood brown, buffy cinnamon, or light tawny-ochraceous
       Spotted-crowned Antvireo, *Dysithamnus puncticeps*

Vol. IV	p. 811: nape, upper back between wood brown and raw umber (equated to light buffy brown)
	Gaumer's Attila, *Attila citreopygus gaumeri*
Vol. IV	p. 762: throat, chest deep wood brown, cinnamon, or grayish cinnamon
	Russet Manakin, *Scotothorus amazonus stenorhynchus*
*Vol. IV	p. 195: rump, tail coverts wood brown (browner than hair brown) (grayer in winter)
	San Lucas Thrasher, *Toxostoma c. cinereum*

Clay Color ~ Color 26

Clay Color is somewhat akin to both Buffy Brown and Fawn Color, although a little paler and brighter. It is shown in both of Ridgway's color guides, but does not occur often in Bulletin 50. It is grouped here with Sayal Brown and Tawny Olive.

The COLOR GUIDE swatch chosen for Clay Brown has a 9.0YR 5.5/5.5 notation.

Notations of Colors Related to Clay Color

Color Name	Ridgway Notation	Hamly Notation	Villalobos Notation
Clay Color	XXIX 17''	10.0 YR 5.4/5.0	O-13-7°
Clay Color, measure of 1912		9.2 YR 5.7/5.8*	
Clay Color, measure of 1886		8.3 YR 6.2/6.5*	
Clay Color, Guide swatch		9.0 YR 5.5/5.5*	
Sayal Brown	XXIX 15''i	8.5 YR 5.2/5.0	OOS-10-5°
Tawny-Olive	XXIX 17''i	10.0 YR 5.2/4.5	O-11-4°
Tawny-Olive, measure of 1912		9.3 YR 5.3/4.8*	

Citations for Clay Color

Vol. VII   p. 76: throat, chest light ochraceous-buff or clay color
Roadrunner, *Geococcyx californianus*
Vol. VII   p. 71: below cinnamon, ochraceous, or clay color to ochraceous-tawny
Rufous-rumped Cuckoo, *Morococcyx e. erythropygus*
Vol. VI   p. 141: nape, crest tawny to clay color or dull ochraceous
Chestnut-colored Woodpecker, *Celus castaneus*
Vol. IV   p. 117: below dull brownish buff or clay color; flanks deep clay color or cinnamon-buff
Gray's Thrush, *Planesticus g. grayi*

Citations for Sayal Brown

Vol. VIII   p. 113: pileum, jugular band sayal brown or dull cinnamon
♀Cinnamomeous Plover, *Pagolla wilsonia cinnamomina*
Vol. VII   p. 361: head, neck sayal brown or cinnamon
Porto Rican Dove, *Zenaida zenaida lucida*

Citations for Tawny-Olive

*Vol. VII   p. 203: hindneck tawny-olive
Red-eared Parrot, *Pyrilia h. haematotis*

Vol. VI	p. 478: side of throat and neck olive tawny or tawny olive-greenish
	Turquoise-browed Motmot, *Eumota s. superciliosa*
Vol. VI	p. 462: belly tawny-olive or rufescent olive tawny
	Small-billed Motmot, *Momotus subrufescens conexus*
Vol. VI	p. 129: wings yellower than tawny-olive (= golden brown)
	♂ Lichtenstein's Woodpecker, *Chloronerpes aeruginosus*
Vol. II	p. 124: above rich brown, between tawny-olive and mummy brown
	♀ White-throated Shrike-Tanager, *Lanio leucothorax*

Drab ~ Color 27

Drab is defined in a dictionary as a brown, with a red yellow hue, of low saturation and medium brilliance. It is a dull, but not extremely dark, yellowish brown with a somewhat grayish cast—a grayish yellowish brown.

The word *drab,* as distinguished from the color, is often used by Ridgway, but he seldom uses the specific color Drab; only three clear-cut, unmodified references are found in Bulletin 50. Occasionally the color is combined with modifying terms that indicate its relationship with other colors, but even this usage is infrequent.

Ridgway seems to find it more agreeable to use two other grayish brown colors more freely: Hair Brown and Broccoli Brown. Each of these is found in Bulletin 50 perhaps fifty times more than Drab. Broccoli Brown, despite its frequent appearance in Bulletin 50, is not shown in *Color Standards* at all. It seems to be of a slightly lighter and richer color than Hair Brown, which is shown in *Color Standards* just below the Drab swatch. It has a darker and grayer quality than Drab.

Some forty colors in *Color Standards* combine *drab* with other color names. Only three of them are of interest in the range of Drab colors: Light Drab, Cinnamon Drab, and Light Brownish Drab (not Brownish Drab). They are nearly synonymous with Drab.

With the exception of Drab, none of the color names in this group has much value for ornithologists. Broccoli Brown and Hair Brown do not convey any clear sense of color. Even Drab may be subject to some question; to many eyes, drabness may connote a much duller quality of color than is depicted in the colors of this group. However, Drab can be used for all the colors of this group, and it can be qualified with words such as *paler, darker, grayer,* etc.

The COLOR GUIDE swatch chosen for Drab has a 9.0YR 5.5/2.5 notation.

Citations for Cinnamon-Drab

*Vol. VII   p. 357: pileum cinnamon-drab, surrounded by fawn color
               Zenaida Dove, *Zenaida z. zenaida*

Citations for Light Brownish Drab

*Vol. VII   p. 451: occiput, nape light brownish drab, with faint bronze gloss
               Buff-bellied Dove, *Leptotila f. fulviventris*

Notations of Colors Related to Drab

Color Name	Ridgway Notation	Hamly Notation	Villalobos Notation
Drab	XLVI 17''''	10.0 YR 5.5/3.0	OOS-10-3°
Drab, measure of 1912		8.8 YR 5.1/2.1*	
Drab, measure of 1886		8.1 YR 5.6/2.5*	
Drab, Guide swatch		9.0 YR 5.5/2.5*	
Light Drab	XLVI 17''''b	10.0 YR 5.8/2.0	OOS-10-3°
Light Brownish Drab	XLV 9''''b	6.0 YR 5.4/2.0	SO-11-3°
Cinnamon Drab	XLVI 13''''	5.0 YR 5.5/2.5	SO-11-3°
Hair Brown	XLVI 17''''i	10.0 YR 4.4/1.0	OOS- 9-2°
Hair Brown, measure of 1912		8.7 YR 4.3/1.3*	
Broccoli Brown, 1886			OOS- 8-3°

Citations for Light Drab

Vol. XI	p. 338: tail light drab to drab Mirador Insect Hawk, *Buteo magnirostris griseocauda*
Vol. VI	p. 287: chin and throat light drab, nearly écru drab ♀Williamson's Woodpecker, *Sphyrapicus thryoideus*
Vol. VI	p. 218: underparts light drab or buffy drab gray Harris' Woodpecker, *Dryobates villosus harrisi*

Citations for Drab

Vol. XI	p. 338: upper parts dark drab to hair brown; tail light drab to drab Mirador Insect Hawk, *Buteo magnirostris griseocauda*
Vol. VIII	p. 140: upper parts deep drab or between drab and hair brown Azara's Ring Plover, *Charadrius collaris*
Vol. VIII	p. 120: above general color between deep drab and hair brown Ringed Plover, *Charadrius hiaticola*
Vol. VIII	p. 32: upper parts deep drab or nearly hair brown American Oystercatcher, *Haematopus p. palliatus*
Vol. VII	p. 398: upper parts deep drab to hair brown Florida Ground Dove, *Chaemepelia p. passerina*
Vol. VII	p. 390: above general color light hair brown to grayish drab ♂Inca Dove, *Scardafella inca*
Vol. VII	p. 370: above deep drab to light hair brown or grayish drab Curaçao Dove, *Zenaidura ruficauda vinaceo-rufa*
Vol. VII	p. 353: upper parts deep drab or dull buffy brown ♂Yucatan Mourning Dove, *Zenaidura yucatanensis*
Vol. VII	p. 341: upper parts deep drab to deep buffy brown ♂Cuban Mourning Dove, *Zenaidura m. macroura*
*Vol. VII	p. 321: above drab, with some paler feather margins ♂Bare-eyed Pigeon, *Crossophthalmus gymnophtalmos*
*Vol. VI	p. 287: sides of head drab; pileum and nape deep drab ♀Williamson's Woodpecker, *Sphyrapicus thyroideus*
Vol. VI	p. 261: above deep broccoli brown or drab, paler in summer ♂Arizona Woodpecker, *Dryobates a. arizonae*

Vol. VI	p. 14: above drab or olive drab (not a color), deeper in winter, grayer in summer
	♂ Flicker, *Colaptes a. auratus*
Vol. V	p. 128: tail deep drab or broccoli brown to sepia
	♂ Spotted Antbird, *Hylophylax naevioides*
Vol. III	p. 507: above brown, varying from prout's to broccoli, or to drab
	Rufous-naped Cactus Wren, *Heleodytes rufinucha*
Vol. III	p. 498: scapulars, rump lighter brown, approaching broccoli or drab
	Louisiana Marsh Wren, *Telmatodytes palustris thryophilus*
*Vol. III	p. 411: back, scapulars, rump (in summer) drab, paler posteriorly
	Alaskan Chickadee, *Penthistes cinctus alascensis*
Vol. III	p. 336: back, scapulars drab or hair brown
	Sumichrast's Jay, *Aphelocoma sumicrasti*
Vol. III	p. 105: general color soft drab (below more grayish drab)
	Bohemian Waxwing, *Ampelis garrulus*
Vol. II	p. 207: head, neck, chest broccoli brown, or drab to clove brown
	♂ Cowbird, *Molothrus a. ater*

Citations for Hair Brown

See citations for Drab (Vol. III, p. 336; Vol. VII, pp. 390, 398; Vol. VIII, pp. 32, 120, 140).

Vol. VII	p. 19: above deep hair brown, between hair brown and fuscous
	Southern Yellow-billed Cuckoo, *Coccyzus americanus julieni*
*Vol. VII	p. 12: above nearest hair brown (with faint bronze gloss)
	Yellow-billed Cuckoo, *Coccyzus a. americanus*
*Vol. V	p. 204: wings grayish hair brown
	Dusky-winged Philydor, *Philydor fuscipennis*
*Vol. V	p. 194: above grayish brown, hair brown
	Costa Rican Gray-breasted Synallaxis, *Synallaxis albescens latitabunda*
Vol. IV	p. 625: pileum hair brown or broccoli brown
	Ash-throated Flycatcher, *Myiarchus c. cinerascens*
*Vol. IV	p. 401: tail deep grayish brown, deep hair brown
	Placid Flycatcher, *Myiopagis p. placens*
*Vol. IV	p. 394: tail deep grayish brown, dark hair brown
	Gray-headed Flycatcher, *Rhynchocyclus cinereiceps*
*Vol. IV	p. 392: tail deep grayish brown, dark hair brown
	Yellow-margined Flycatcher, *Rhynchocyclus marginatus*
*Vol. IV	p. 303: wings (partially) deep grayish brown, hair brown (in summer)
	♂ Shore Lark, *Octocoris a. alpestris*
*Vol. IV	p. 259: above light grayish brown, hair brown (grayer in fall)
	Sage Thrasher, *Oroscoptes montanus*
Vol. IV	p. 241: above brownish gray, between hair brown and drab gray
	Gundlach's Mockingbird, *Mimus g. gundlachi*

Vol. IV	p. 201: above hair brown, nearly broccoli, browner than drab gray
	Mazatlan Thrasher, *Toxostoma curvirostre occidentale*
Vol. IV	p. 197: above light grayish brown, light hair brown or between hair brown and broccoli brown (grayer in winter)
	Bendire's Thrasher, *Toxostoma bendirei*
Vol. IV	p. 195: above hair brown, or between hair brown and broccoli (grayer in winter)
	San Lucas Thrasher, *Toxostoma c. cinereum*
Vol. IV	p. 163: above varying drab gray or smoke gray to hair brown
	Townsend's Solitaire, *Myadestes townsendi*
Vol. IV	p. 39: above hair brown or to between hair brown and broccoli
	Kadiak Dwarf Thrush, *Hylocichla g. guttata*
*Vol. IV	p. 15: above grayish brown, deep hair brown
	Red-spotted Bluethroat, *Cyanosylvia suecica*
Vol. III	p. 514: pileum light grayish brown, broccoli to hair brown
	Gray Cactus Wren, *Heleodytes nelsoni*
Vol. III	p. 136: nape, upper back hair brown or pale drab gray
	Grand Cayman Vireo, *Vireosylva caymanensis*
Vol. III	p. 66: above dark hair brown to sepia
	Coban Swallow, *Notiochelidon pileata*
*Vol. III	p. 58: above grayish brown, deep hair brown
	Rough-winged Swallow, *Stelgidopteryx serripennis*
Vol. I	p. 575: above olive brown or olivaceous hair brown
	♀ Morellet's Seedeater, *Sporophila morelleti*

Citations for Broccoli Brown

Broccoli Brown was omitted from *Color Standards* by Ridgway. See citations for Drab (Vol. III, pp. 498, 507) and Hair Brown (Vol. III, p. 514; Vol. IV, pp. 39, 195, 197, 201, 625).

*Vol. VI	p. 382: pileum, nape, upper back deep broccoli brown
	Russet-throated Puffbird, *Hypnelus r. ruficollis*
Vol. V	p. 135: upper parts between broccoli brown and raw umber
	McLeannan's Antthrush, *Phaenostictus m. mcleannani*
Vol. V	p. 128: tail deep drab, or broccoli brown to sepia
	♂ Spotted Antbird, *Hylophylax naevioides*
*Vol. IV	p. 621: pileum broccoli brown
	Mexican Crested Flycatcher, *Myiarchus m. mexicanus*
Vol. IV	p. 287: above deep broccoli brown, to nearly prout's or van-dyke
	Scarlet-cheeked Weaver-Finch, *Estrilda melpoda*
*Vol. IV	p. 216: above grayish brown, deep broccoli brown
	Socorro Thrasher, *Mimodes graysoni*
*Vol. IV	p. 203: above deep grayish brown, dark broccoli brown
	California Thrasher, *Toxostoma redivivum*

Olive-Brown ~ Color 28

Olive-Brown, a hyphenated color name, occurs with great frequency in Bulletin 50. Ridgway relates it closely to the Raw Umber group. Fifteen of over thirty citations for Olive-Brown refer to Raw Umber. Seven of them also refer to Bister and Sepia, which are grouped here with Olive-Brown. They are not equivalent colors, of course, but they are close enough to make it possible to omit them from the NATURALIST'S COLOR GUIDE.

Palmer (1962) shows a swatch for Sepia; it can be used by anyone who may desire to use that color name. In the glossy finish, as published, it measures a 9.1YR 2.9/3.6 notation, perhaps a little too bright when compared with Ridgway's swatch. However, the matte finish is substantially closer, with a 9.4YR 3.9/2.6 notation. But it is not quite equivalent to Olive-Brown.

Olive-Brown is visibly more olivaceous than Raw Umber. It is followed by two additional colors, Brownish Olive and Olive, which are more difficult to separate visually. All three have equivalent value and chroma notations, identified by different hue notations.

The COLOR GUIDE swatch chosen for Olive-Brown has a 10.0YR 4.0/2.0 notation.

Notations of Colors Related to Olive-Brown

Color Name	Ridgway Notation	Hamly Notation	Villalobos Notation
Olive-Brown	XL 17'''k	10.0 YR 3.8/2.0	O-6-3°
Olive-Brown, measure of 1912		8.2 YR 3.7/2.0*	
Olive-Brown, Guide swatch		10.0 YR 4.0/2.0*	
Bister	XXIX 15''m	9.0 YR 3.5/2.0	OOS-5-4°
Bister, measure of 1912		7.6 YR 3.4/2.5*	
Bister, measure of 1886		8.8 YR 3.9/3.1*	
Sepia	XXIX 17''m	1.0 Y 3.4/2.0	OOS-5-4°
Sepia, measure of 1912		10.0 YR 3.5/2.4*	
Sepia, measure of 1886		9.0 YR 4.1/2.8*	
Sepia, Palmer glossy		9.1 YR 2.9/3.6*	OOS-5-4°
Sepia, Palmer matte		9.4 YR 3.9/2.6*	SO-4-2°*

Citations for Olive-Brown

*Vol. VII p. 460: above olive-brown
　　　　　　Rufous-naped Dove, *Leptotila rufinucha*
*Vol. VII p. 71: above olive-brown (back slightly browner)
　　　　　　Rufous-rumped Cuckoo, *Morococcyx e. erythropygus*
Vol. V p. 285: above olive-brown or raw umber
　　　　　Cherrie's Deconychura, *Deconychura typica*
Vol. V p. 254: above olive-brown, nearly raw umber
　　　　　Spotted Woodhewer, *Xiphorhynchus erythropygius*
Vol. V p. 250: above olive-brown, between raw umber and mummy brown
　　　　　Lawrence's Woodhewer, *Xiphorhynchus n. nanus*
Vol. V p. 233: above olive-brown, between bister and raw umber
　　　　　Costa Rican Woodhewer, *Dendrocolaptes validus costaricensis*
*Vol. V p. 232: above olive-brown (called fulvous by Sclater in 1890)
　　　　　Spotted-necked Woodhewer, *Dendrocolaptes puncticollis*
Vol. V p. 229: above olive-brown, nearly raw umber to mars brown
　　　　　Barred Woodhewer, *Dendrocolaptes s. sancti-thomae*
Vol. V p. 185: above olive-brown, between sepia and raw umber
　　　　　Lawrence's Spinetail, *Acrorchilus erythrops rufigenis*
Vol. V p. 135: above general color light olive-brown, between broccoli brown and raw umber
　　　　　McLeannan's Antthrush, *Phaenostictus m. mcleannani*
Vol. V p. 121: above general color light olive-brown, between broccoli brown and isabella, brightening to raw umber on coverts
　　　　　Yucatan Antthrush, *Formicarius moniliger pallidus*
Vol. V p. 85: pileum, nape olive-brown, between prout's and raw umber
　　　　　Northern Long-billed Antwren, *Ramphocaenus r. rufiventris*
*Vol. V p. 50: upper parts general color olive-brown
　　　　　♀Slaty Antshrike, *Erionotus punctatus atrinucha*
Vol. IV p. 804: lower back, scapulars olive-brown or bister
　　　　　Gray-headed Attila, *Attila tephrocephalus*
Vol. IV p. 672: above olive-brown or brownish olive to bister
　　　　　Derby Flycatcher, *Pitangus sulphuratus derbianus*
Vol. IV p. 591: deep olive-brown or raw umber on "young" specimen
　　　　　Ruddy Flycatcher, *Empidonax fulvifrons rubicundus*
Vol. IV p. 498: pileum deep olive-brown, bister, or sepia
　　　　　Dusky-tailed Flycatcher, *Mitrephanes p. phaeocercus*
*Vol. IV p. 484: above olive-brown
　　　　　Brown Flycatcher, *Cnipodectes subbrunneus*
*Vol. IV p. 354: upper parts olive-brown
　　　　　♂Mexican Royal Flycatcher, *Onychorhynchus m. mexicanus*
*Vol. IV p. 166: back and rump olive-brown
　　　　　Brown-backed Solitaire, *Myadestes o. obscurus*
Vol. IV p. 115: above olive-brown, between bister and raw umber
　　　　　Black-billed thrush, *Planesticus nigrirostris*
*Vol. IV p. 74: above olive-brown
　　　　　Guadeloupe Forest Thrush, *Cichlherminia h. herminieri*

Vol. IV	p. 52: above olive-brown, varies from hair brown to between broccoli and raw umber (summer); olive-brown to russet or between russet and raw umber (winter)
	Russet-backed Thrush, *Hylocichla u. ustulata*
Vol. IV	p. 28: above olive-brown or raw umber
	Frantzius' Nightingale Thrush, *Catharus f. frantzii*
Vol. III	p. 628: back varies from olive-brown to nearly mummy brown
	Stripe-breasted Wren, *Thryophilus thoracicus*
Vol. III	p. 534: above light olive-brown or raw umber
	Spot-breasted Wren, *Pheugopedius m. maculipectus*
*Vol. III	p. 533: above olive-brown
	Tawny-bellied Wren, *Pheugopedius hyperythrus*
Vol. III	p. 426: above brownish olive or olive-brown, rump browner
	Black-eared Bush-Tit, *Psaltriparus m. melanotis*
Vol. III	p. 218: above olive-brown or russet
	Tawny-crowned Greenlet, *Pachysilvia o. ochraceiceps*
*Vol. II	p. 436: tail olive-brown
	Swainson's Warbler, *Helinaia swainsonii*
Vol. II	p. 125: back olive-brown or deep raw umber
	Black-rumped Shrike-Tanager, *Lanio melanopygius*
Vol. I	p. 575: above olive-brown or olivaceous hair brown
	Morellet's Seedeater, *Sporophila morelleti*

Citations for Bister

See citations for Olive-Brown (Vol. V, p. 233).

Vol. VI	p. 624: above general color bister to nearly clove brown
	Richardson's Owl, *Cryptoglaux tengmalmi richardsoni*
Vol. VI	p. 506: above general color between bister and mummy
	♂ Chuck-will's-widow, *Antrostomus carolinsensis*
*Vol. V	p. 172: pileum brown, nearly bister
	Mexican Xenops, *Xenops genibarbis mexicanus*
Vol. IV	p. 762: above light bister, pileum ruddier, rump paler
	Russet Manakin, *Scotothorus amazonus stenorhynchus*
Vol. IV	p. 285: above between bister and broccoli brown
	♂ "young" Hooded Weaver-Finch, *Spermestes cucullata*
Vol. IV	p. 24: above deep brownish olive, bister
	Sooty Nightingale Thrush, *Catharus mexicanus fumosus*
Vol. IV	p. 22: tail and coverts inclining to bister
	Black-headed Nightingale Thrush, *Catharus m. mexicanus*
Vol. III	p. 593: above between prout's and sepia or bister
	Martinique Wren, *Troglodytes martinicensis*

Citations for Sepia

Note the frequent use of the term sooty brown here; also see the citations for Clove Brown and Olive-Brown (Vol. V, p. 185).

Vol. X	p. 23: back, rump dull dusky sepia to mummy brown
	Northern Crested Guan, *Penelope p. purpurascens*
Vol. VI	p. 261: pileum, nape deep sooty brown, warm to dark sepia
	♂ Arizona Woodpecker, *Dryobates a. arizonae*

Vol. V	p. 694: pileum dark sooty brown, sepia to near clove brown
	Jamaican Palm Swift, *Tachornis p. phoenicobia*
Vol. V	p. 264: pileum sepia to light sepia
	Streaked Woodhewer, *Picolaptes l. lineaticeps*
Vol. V	p. 261: pileum, nape light sepia to dark raw umber
	Allied Woodhewer, *Picolaptes a. affinis*
Vol. IV	p. 849: pileum, nape prout's to between vandyke and sepia
	♂ Jamaican Becard, *Platypsaris niger*
Vol. IV	p. 652: pileum deep sooty brown, dark sepia to near clove brown
	Sad Flycatcher, *Myiarchus barbirostris*
*Vol. IV	p. 642: pileum deep sooty brown, dark sepia
	Lawrence's Flycatcher, *Myiarchus l. lawrenceii*
Vol. IV	p. 638: pileum dark sooty brown, dark sepia to near clove brown
	Porto Rican Flycatcher, *Myiarchus antillarum*
Vol. IV	p. 634: pileum deep sooty brown, sepia to near clove brown
	Santo Domingo Flycatcher, *Myiarchus dominicensis*
Vol. IV	p. 633; pileum deep sooty brown, dark sepia to near clove brown
	Stolid Flycatcher, *Myiarchus stolidus*
*Vol. IV	p. 632: pileum sepia
	Yucatan Flycatcher, *Myiarchus yucatanensis*
Vol. IV	p. 606: above light olive-brown to grayish sepia
	Galapagos Flycatcher, *Eribates magnirostris*
Vol. IV	p. 136: above dark sooty brown, between sepia and clove brown
	♂ Aztec Thrush, *Ridgwayia pinicola*
Vol. III	p. 593: above between prout's brown and sepia or bister
	Martinique Wren, *Troglodytes martinicensis*
Vol. III	p. 503: upper parts sooty brown, deep sepia
	♂ White-headed Cactus Wren, *Heleodytes albobrunneus*
*Vol. II	p. 124: pileum sepia
	White-throated Shrike-Tanager, *Lanio leucothorax*

Brownish Olive ~ Color 29

Brownish Olive is discussed in the Introduction, where numerous tests of the plumage of backs of the Great Crested Flycatcher, *Myiarchus crinitus,* are described. The tests indicated that a distinction could and should be made between the color of spring versus autumn plumages. Spring plumage is duller olive and should be called Brownish Olive, but autumn plumage is more clearly plain Olive. The differences are so slight, however, that one might question whether the specific color name Brownish Olive is warranted. But I have retained the name and color to conform with Ridgway and Palmer. Also, the tests showed that the spring plumage could not be called Olive-Brown. The descriptions of this flycatcher by various ornithologists are in utter disagreement: Ridgway says it is olive above; Blake says dull olive; Land says brownish olive; Bent says olive-brown; Imhof says olive-green.

Palmer (1962) shows a swatch for Brownish Olive and one for Olive. They are both somewhat more brown than the colors shown in the NATURALIST'S COLOR GUIDE, the former trending toward Olive-Brown, the latter closer to Brownish Olive, but not Olive. The differences between Palmer and the COLOR GUIDE are merely a matter of choice, or possibly of printing accuracy. Measurement data for all these swatches are included in the tabulation.

The COLOR GUIDE swatch chosen for Brownish Olive has a 2.5Y 4.0/2.0 notation.

Notations of Colors Related to Brownish Olive

Color Name	Ridgway Notation	Hamly Notation	Villalobos Notation
Brownish Olive	XXX 19″m	2.5 Y 3.8/3.0	O-5-2°
Brownish Olive, measure of 1912		2.6 Y 2.6/2.0*	
Brownish Olive, Palmer glossy		1.6 Y 3.5/2.5*	
Brownish Olive, Palmer matte		2.0 Y 4.0/1.7*	O-6-3°
Brownish Olive, Guide swatch		2.5 Y 4.0/2.0*	
Olive, Palmer glossy		2.0 Y 3.2/3.0*	OOY-5-2°
Olive, Palmer matte		2.2 Y 3.7/1.7*	

Citations for Brownish Olive

Certain specimens were measured visually, resulting in the following notations: *Myiarchus crinitus,* spring plumage, 2.7Y 3.9/2.1 and 2.2Y 3.8/2.0; *Myiarchus tuberculifer* 1.9Y 3.7/1.9; *Myiarchus yucatanensis* 2.0Y 3.8/2.0; *Dysithamnus mentalis septentrionalis* 2.8Y 3.7/1.8; *Elaenia f. flavogaster* 2.9Y 3.8/2.2 and 2.7Y 3.8/2.0.

*Vol. V	p. 145: above brownish olive
	Costa Rican Grallaricula, *Grallaricula costaricensis*
Vol. IV	p. 759: slightly brownish olive to clear olive
	Olive Manakin, *Scotothorus verae-pacis dumicola*
Vol. IV	p. 647: pileum brownish olive or light sepia
	Querulous Flycatcher, *Myiarchus lawrencii querulus*
Vol. IV	p. 584: above brownish olive or olive-brown
	White-throated Flycatcher, *Empidonax albigularis*
Vol. IV	p. 576: above brownish olive, varying to grayish olive
	Western Flycatcher, *Empidonax d. difficilis*
*Vol. IV	p. 561: above brownish olive
	Least Flycatcher, *Empidonax minimus*
*Vol. IV	p. 536: above brownish olive (head darker, tail lighter)
	Jamaican Wood Peewee, *Blacicus pallidus*
Vol. IV	p. 498: above lighter or less brownish olive
	Dusky-tailed Flycatcher, *Mitrephanes p. phaeocercus*
*Vol. IV	p. 434: above brownish olive
	Frantzius' Elaenia, *Elaenia f. frantzii*
*Vol. IV	p. 384: above brownish olive
	♂ White-throated Spade-billed Flycatcher, *Platyrinchus albogularis*
Vol. IV	p. 382: above brownish olive or olive-brown
	♂ Mexican Spade-billed Flycatcher, *Platyrinchus cancrominus*
*Vol. III	p. 686: above brownish olive
	Wren-Tit, *Chamaea f. fasciata*
Vol. III	p. 426: above brownish olive or olive-brown
	♂ Black-eared Bush-Tit, *Psaltriparus m. melanotis*
Vol. III	p. 210: above brownish olive or olive-brown
	Latimer's Vireo, *Vireo latimeri*

Olive ~ Color 30

Olive as a specific color is fully discussed in the Introduction. It is the ornithologist's most important, most used, perhaps most abused color. *Color Standards* includes some three dozen colors of an olivaceous nature, each with its own name and representative swatch. One might question whether such an extensive quantity of olive colors is essential.

In the NATURALIST'S COLOR GUIDE, only nine olivaceous colors are shown, each a unique, distinct color with a narrow range of difference. The borderlines between them sometimes overlap, and perhaps even fewer would be adequate.

When tested under daylight lamps, olive fruits were all too bright yellowish or greenish a hue to be equated with Ridgway's Olive. Cocktail olives measured about 5.5Y 5.8/5.6; bottled olives ranged from 5.5Y 4.7/4.3 to 6.0Y 5.7/7.1; only when fresh olives were allowed to dry out substantially did they even approach Ridgway's Olive, then ranging from 6.5Y 3.4/2.7 to 6.5Y 3.4/1.9—still displaying a richer quality of color (black olives were obviously omitted from these measurements).

The COLOR GUIDE swatch chosen for Olive has a 4.5Y 4.0/2.0 notation.

Notations of Colors Related to Olive

Color Name	Ridgway Notation	Hamly Notation	Villalobos Notation
Olive	XXX 21''m	5.0 Y 3.8/3.0	OOY-5-2°
Olive, measure of 1912		4.6 Y 3.6/1.8*	
Olive, Guide swatch		4.5 Y 4.0/2.0*	
Deep Olive	XL 21'''k	5.0 Y 4.4/3.0	OY-6-3°
Dark Olive	XL 21'''m	5.0 Y 3.5/2.0	OOY-3-2°

Citations for Olive

*Vol. VII p. 55: above olive
Jamaican Hyetornis, *Hyetornis pluvialis*

Vol. V p. 280: pileum and nape grayish olive; back and scapulars russet, sometimes tinged with olive, but the Venezuelan subspecies (*Sittasomus g. griseus*) shows upper parts much more uniformly olive
Mexican Sittasomus, *Sittasomus s. sylvioides*

*Vol. V	p. 156: above olive, with only few buff streaks
	Lizano's Antpitta, *Hylopezus perspicillatus lizanoi*
*Vol. V	p. 155: above olive, but scapulars with streaks of buff
	Lawrence's Antpitta, *Hylopezus p. perspicillatus*
Vol. V	p. 148: above olive, but feathers with black margins
	Guatemalan Antpitta, *Grallaria g. guatimalensis*
Vol. V	p. 59: above olive or grayish olive (♀plain olive)
	Streaked-crowned Antvireo, *Dysithamnus striaticeps*
Vol. V	p. 56: above olive (varying from grayish to brownish olive)
	♀Northern Antvireo, *Dysithamnus mentalis septentrionalis*
*Vol. IV	p. 759: upper parts clear olive to slightly brownish olive
	Olive Manakin, *Scotothorus verae-pacis dumicola*
*Vol. IV	p. 675: above olive
	Lictor Flycatcher, *Pitangus lictor*
*Vol. IV	p. 661: above including pileum, olive general color
	Insolent Flycatcher, *Myiodynastes maculatus insolens*
Vol. IV	p. 659: above (not head) olive or light grayish brown
	Noble Flycatcher, *Myiodynastes maculatus nobilis*
Vol. IV	p. 656: above light olive usually tinged with buffy
	Sulphur-bellied Flycatcher, *Myiodynastes luteiventris*
*Vol. IV	p. 652: above olive
	Sad Flycatcher, *Myiarchus barbirostris*
*Vol. IV	p. 642: above olive
	Lawrence's Flycatcher, *Myiarchus l. lawrenceii*
*Vol. IV	p. 632: above olive
	Yucatan Flycatcher, *Myiarchus yucatanensis*
*Vol. IV	p. 619: above usually decidedly olive
	Grenada Flycatcher, *Myiarchus oberi nugator*
Vol. IV	p. 617: above deep to dark olive
	Ober's Flycatcher, *Myiarchus o. oberi*
*Vol. IV	p. 613: above olive
	Crested Flycatcher, *Myiarchus crinitus*
Vol. IV	p. 586: above olive or olive-brown, varying much in intensity
	Black-capped Flycatcher, *Empidonax atriceps*
*Vol. IV	p. 575: above olive in spring (deeper in winter)
	Chancol Flycatcher, *Empidonax trepidus*
*Vol. IV	p. 567: above olive
	Wright's Flycatcher, *Empidonax wrightii*
Vol. IV	p. 555: above olive, but varying grayish to brownish hues
	Traill's Flycatcher, *Empidonax t. traillii*
Vol. IV	p. 540: above dark olive
	Grenada Wood Pewee, *Blacicus flaviventris*
Vol. IV	p. 537: above deep olive
	Porto Rico Wood Pewee, *Blacicus ploncol*
Vol. IV	p. 535: above deep olive, rump lighter
	Haitian Wood Pewee, *Blacicus hispaniolensis*
*Vol. IV	p. 526: above olive
	Short-legged Wood Pewee, *Myiochanes brachytarsus*
*Vol. IV	p. 518: above olive; pileum darker, rump lighter
	Wood Pewee, *Myiochanes virens*

*Vol. IV	p. 492: above olive
	Salvin's Flycatcher, *Aphanotriccus capitalis*
*Vol. IV	p. 444: above olive
	Cayenne Flycatcher, *Myiozetetes c. cayenensis*
*Vol. IV	p. 436: above olive
	Jamaican Elaenia, *Elaenia fallax*
*Vol. IV	p. 429: above olive
	Northern Elaenia, *Elaenia martinica subpagana*
Vol. IV	p. 426: above olive, varying grayish olive to olive-brown
	Antillean Elaenia, *Elaenia m. martinica*
*Vol. IV	p. 379: above olive
	♂ Lawrence's Spade-billed Flycatcher, *Placostomus superciliaris*
Vol. IV	p. 113: above olive or grayish olive
	Naked-eyed Thrush, *Planesticus gymnophthalmus*
Vol. IV	p. 59: above olive to grayish olive in summer; brighter, olive to olive-sepia in winter
	Gray-cheeked Thrush, *Hylocichla a. aliciae*
Vol. IV	p. 55: above olive to grayish olive in summer; deep olive to brownish olive in winter
	Olive-backed (Swainson's) Thrush, *Hylocichla ustulata swainsonii*
*Vol. IV	p. 22: above olive
	Black-headed Nightingale Thrush, *Catharus m. mexicanus*
*Vol. III	p. 195: above olive, trending to greenish olive posteriorily
	Hutton's Vireo, *Vireo h. huttoni*
*Vol. II	p. 642: above olive
	Water-Thrush, *Seiurus n. noveboracensis*
*Vol. II	p. 541: above olive (spring and summer)
	♀ Black-throated Blue Warbler, *Dendroica c. caerulescens*
*Vol. II	p. 537: above olive (spring and summer)
	♀ Cape May Warbler, *Dendroica tigrina*
*Vol. II	p. 436: above olive
	Swainson's Warbler, *Helenaia swainsonii*
*Vol. II	p. 380: above olive, rump lighter
	♀ Mexican Diglossa, *Diglossa baritula*
*Vol. II	p. 156: above olive
	Porto Rican Tanager, *Nesospingus speculiferus*
Vol. II	p. 123: pileum, nape dark olive
	♀ Mexican Shrike-Tanager, *Lanio aurantius*
*Vol. II	p. 72: above always olive
	Cozumel Spindalis, *Spindalis benedicti*
*Vol. II	p. 69: above, including head, olive
	♀ Cuban Spindalis, *Spindalis pretrei*
Vol. II	p. 60: back olive (with bluish gloss in certain lights)
	Abbot Tanager, *Tanagra abbas*
*Vol. I	p. 571: above olive
	♀ Black Seedeater, *Sporophila corvina*
Vol. I	p. 569: above olive (rather light)
	♀ Yellow-bellied Seedeater, *Sporophila gutturalis*
Vol. I	p. 554: above uniform dark olive
	♀ Martinique Bullfinch, *Pyrrhulagra n. noctis*

86

Vol. I	p. 527: above olive or olive brownish
	♀ Blue-black Grassquit, *Volatina jacarini splendens*
Vol. I	p. 461: above dusky olive
	Yellow-throated Sparrow, *Atlapetes gutturalis*
*Vol. I	p. 39: above, limited to nape and back, olive
	♂ Evening Grosbeak, *Hesperiphona v. vespertina*

Maroon ~ Color 31

Maroon is included in the NATURALIST'S COLOR GUIDE for various reasons. It fills a gap in the very dark reddish yellow area of this group of colors. It is a commonly used color name, although some people may think of it as having a more vivid quality than that depicted by Ridgway. It has a strongly brownish tone that relates it but does not equate it to Chestnut and Bay. It is defined in a dictionary as: "formerly the color chestnut; of a yellowish red hue."

Ridgway depicts Maroon in both *Color Standards* and *Nomenclature*. However, he rarely uses it in Bulletin 50, and the two citations listed are not very helpful. One citation indicates that he thought of it as dusky red, but the species so described is only tinged with the color and is only sometimes wholly maroon. The other citation describes the plumage as "orange-maroon or madder brown." Also included is a citation saying "crimson-maroon." Neither orange-maroon nor crimson-maroon is a specific color. Madder Brown is a specific color of a very different character. Bonaparte's Tanager may be the best plumage available to represent the color.

The tabulation includes other Ridgway colors, based partly on their resemblance to each other in *Color Standards* and partly on the similarity of the notations given them by Villalobos. By and large they appear to be too grayish or brownish, some excessively dark in tone, to be considered truly in the Maroon range.

The COLOR GUIDE swatch chosen for Maroon has a 7.5R 2.5/5.0 notation, which is a more vivid color than Ridgway's. Hues of the 5.0R area can be included in the range of Maroon because dark colors are not greatly influenced by hue alone.

Notations of Colors Related to Maroon

Color Name	Ridgway Notation	Hamly Notation	Villalobos Notation
Maroon	I 3m	7.5 R 2.4/4.0*	RS-6-3
Maroon, Guide swatch		7.5 R 2.5/5.0*	
Haematite Red	XXVII 5″m	7.5 R 3.4/2.0	RS-6-3°
Victoria Lake	I 1m	5.0 R 2.0/4.0	RS-4-6°
Indian Red	XXVII 3″k	5.0 R 4.0/4.0	RS-7-4°
Dark Indian Red	XXVII 3″m	5.0 R 3.8/2.5	RS-6-3°
Mineral Red	XXVII 1″k	2.5 R 3.6/4.0	R-7-4°
Vandyke Red	XIII 1′k	2.5 R 3.8/5.0	RS-7-5°

Citations for Maroon

Vol. II   p. 116: head, back dark crimson-maroon
          ♂ Crimson-backed Tanager, *Ramphocelus d. dimidiatus*

Vol. II   p. 115: head, back, chest sometimes wholly maroon, but usually only tinged dusky red or maroon
          ♂ Bonaparte's Tanager, *Ramphocelus luciani*

Vol. II   p. 46: head bright glossy orange-maroon or madder brown
          ♂ Lavinia's Tanager, *Calospize lavinia*

Chestnut ~ Color 32

Chestnut is combined with Bay, Auburn, and Brick Red, despite visual differences in hue. Chestnut is frequently used in Bulletin 50; Bay, Auburn, and Brick Red are seldom used. There should not be any serious loss in their elimination as colors if Chestnut is used as the basic color and modifiers are added when needed. Brick Red, the ruddiest of the group, is a very indefinite name; bricks occur in many different reddish colors.

In the tabulation, the measurements recorded by Villalobos are practically identical. The Hamly correlations are close but not identical; they indicate more accurately the minor differences.

The Palmer swatch for Chestnut is a good color. It has a notation of 1.0YR 3.0/4.2, which is within the range of the Chestnut group.

Specimens of some species cited under Chestnut were measured spectrophotometrically with the following results. The Guatemalan Ruddy Rail or Ruddy Crake (the upper back and collar areas), described by Ridgway as bright chestnut, measured within the Burnt Umber group, including Chestnut-Brown. If related to Chestnut, the area would be considered dark rather than bright, although it is brighter than the more chocolate-colored areas of the bird's adjacent back and scapulars.

The upper parts of the Sclater's Antbird, described by Ridgway as deep chestnut, measured within the Fuscous group, including Bone Brown, which is decidedly darker and less ruddy than Chestnut.

The upper parts of the female Cherrie's Antbird, described as "more tawny chestnut" than the plain chestnut back of the male bird, measured closer to Russet than Chestnut and lacked the more intense chroma of Tawny.

The lower rump and tail of the Buff-throated Foliage-Gleaner, described as chestnut, measured within the Burnt Umber group, specifically as Vandyke Brown. This would indicate a color darker and less ruddy than Chestnut, while still somewhat more ruddy than the Bone Brown found for the Sclater's Antbird.

The above examples indicate a wide range of disagreement between verbal descriptions and scientific measurements. Perhaps the latter should also be questioned; perhaps the diffuse reflectances from plumages do not register the same notations as derived from a uniformly flat surface of the same color. However, when they are cross-checked by visual comparisons between the plumages and a large file of Munsell swatches, the scientific measurements are acceptable.

90

The COLOR GUIDE swatch chosen for Chestnut has a 1.0YR 3.0/5.0 notation, which is close to Palmer's.

Notations of Colors Related to Chestnut

Color Name	Ridgway Notation	Hamly Notation	Villalobos Notation
Chestnut	II 9m	10.0 R 3.0/5.0	SSO-5-6°
Chestnut, measure of 1912		2.9 YR 3.3/5.1*	
Chestnut, measure of 1886		3.9 YR 4.2/5.3*	
Chestnut, Palmer glossy		1.0 YR 3.0/4.2*	SSO-5-6°
Chestnut, Palmer matte		1.0 YR 3.5/3.2*	
Chestnut, Guide swatch		1.0 YR 3.0/5.0*	
Bay	II 7m	10.0 R 2.6/6.0	SSO-5-6°
Bay, measure of 1912		1.5 YR 3.2/4.0*	
Auburn	II 11m	2.5 YR 3.4/4.0	SO-5-5°
Auburn, measure of 1912		4.3 YR 3.4/3.8*	
Brick Red	XIII 5'k	10.0 R 3.6/5.0	SSO-6-5°
Hay's Russet	XIV 7'k	10.0 R 4.0/5.0	SSO-7-6°

Citations for Chestnut

Vol. XI   p. 671: belly, thighs deep bright hazel but generalized as chestnut
Orange-breasted Falcon, *Falco deiroleucus*

Vol. XI   p. 675: belly, thighs deep hazel to deep cinnamon-rufous, but generalized as chestnut
Bat Falcon, *Falco albigularis*

Vol. XI   p. 163: thighs between russet and chestnut
Bicolored Hawk, *Accipiter b. bicolor*

Vol. IX   pp. 168, 169: upper back, sides of neck bright chestnut; below cinnamon, heavily suffused with deep chestnut (specimen 45686 measured back 6.5YR 1.8/2.0 and collar 4.5YR 2.0/3.5)
Guatemalan Ruddy Rail, *Laterallus r. ruber*

Vol. VIII   p. 431: lower neck rich dark chestnut (continuing backward)
♀Wilson's Phalarope, *Steganopus tricolor*

Vol. VIII   p. 418: underparts deep purplish cinnamon or vinaceous-brown, but called deep chestnut by Blake
♀Red Phalarope, *Phalaropus fulicarius*

Vol. VII   p. 484: back deep chestnut or chestnut-brown; other parts above dull chestnut or bay or chestnut-brown
♂White-bellied Quail-Dove, *Oreopelia violacea albiventer*

Vol. VII   p. 478: above general color chestnut or rufous chestnut
♂Ruddy Quail-Dove, *Oreopelia montana*

Vol. VII   p. 56: below, except belly, chestnut or bay
Haitian Hyetornis, *Hyetornis rufigularis*

Vol. VII   p. 48: above chestnut, deepening to bay on tail, fading to rufous chestnut on upper back and nape, tail bright bay
Central American Squirrel Cuckoo, *Piaya cayana thermophila*

91

Vol. VI	p. 466: pileum deep chestnut (nape chestnut tawny)
	♂ Chestnut-headed Motmot, *Momotus castaneiceps*
Vol. VI	p. 357: under tail coverts chestnut or deep cinnamon-rufous
	Blue-throated Toucanet, *Aulacorhynchus c. caeruleogularis*
Vol. VI	p. 355: under tail coverts chestnut or deep cinnamon-rufous
	Emerald Toucanet, *Aulachorhynchus p. prasinus*
Vol. VI	p. 354: under tail coverts light chestnut or deep cinnamon-rufous
	Wagler's Toucanet, *Aulachorhynchus wagleri*
*Vol. VI	p. 350: thighs chestnut
	Cassin's Araçari, *Selenidera spectabilis*
*Vol. VI	p. 347: thighs chestnut
	Red-rumped Araçari, *Pteroglossus sanguineus*
Vol. VI	p. 345: nuchal collar dark chestnut; thighs chestnut or deep cinnamon-rufous
	Frantzius' Araçari, *Pteroglossus frantzii*
Vol. VI	p. 342: nuchal collar, thighs chestnut or deep cinnamon-rufous
	Collared Araçari, *Pteroglossus t. torquatus*
Vol. VI	p. 141: body, wings chestnut or rufous chestnut, plus barring
	Chestnut-colored Woodpecker, *Celeus castaneus*
*Vol. V	p. 420: tail chestnut
	♂ Bang's Hummingbird, *Amizilis bangsi*
*Vol. V	p. 419: tail chestnut (specimen 394038 measured 3.9YR 2.3/3.1)
	Coral-billed Hummingbird, *Amizilis rutila corallirostris*
Vol. V	p. 293: above chestnut or rufous chestnut, back dullish
	Ruddy Dendrocincla, *Dendrocincla h. homochroa*
*Vol. V	p. 285: rump, tail chestnut
	Cherrie's Deconychura, *Deconychura typica*
*Vol. V	p. 242: tail deep chestnut
	Black-striped Woodhewer, *Xiphorhynchus l. lachrymosus*
Vol. V	p. 229: tail chestnut or cinnamon-rufous
	Barred Woodhewer, *Dendrocolaptes s. sancti-thomae*
*Vol. V	p. 220: tail chestnut
	Pale-throated Automolus, *Automolus p. pallidigularis*
*Vol. V	p. 217: rump, tail chestnut
	Buff-throated Automolus, *Automolus c. cervinigularis*
*Vol. V	p. 204: above chestnut
	Dusky-winged Philydor, *Philydor fuscipennis*
*Vol. V	p. 166: rump chestnut
	Gray-throated Sclerurus, *Sclerurus canigularis*
*Vol. V	p. 113: above ♂ chestnut, ♀ more tawny chestnut (specimen 182765 measured 6.5YR 2.0/2.7)
	♂ Cherrie's Antbird, *Myrmeciza exsul occidentalis*
Vol. V	p. 111: above deep chestnut (specimen 246735 measured 6.0YR 1.5/2.2)
	Sclater's Antbird, *Myrmeciza e. exsul*
Vol. V	p. 40: pileum bright chestnut or rufous chestnut
	♀ Mexican Antshrike, *Thamnophilus doliatus mexicanus*
*Vol. V	p. 29: above chestnut (pileum darker, rump duller)
	♀ Holland's Antshrike, *Taraba t. transandeana*
*Vol. IV	p. 218: under tail coverts chestnut
	Catbird, *Galeoscoptes carolinensis*

*Vol. IV	p. 152: back chestnut
	♂ Chestnut-backed Bluebird, *Sialia mexicana bairdi*
*Vol. IV	p. 148: chest, sides chestnut (spring and summer)
	♂ Mexican Bluebird, *Sialia m. mexicana*
Vol. III	p. 624: above bright chestnut
	Bay Wren, *Thryophilus c. castaneus*
*Vol. III	p. 532: back chestnut
	Black-bellied Wren, *Pheugopedius fasciato-ventris melanogaster*
Vol. III	p. 505: hindneck chestnut or dark cinnamon-rufous; back dull chestnut or cinnamon-rufous (summer); dull chestnut or cinnamon-chestnut (winter)
	Hooded Cactus Wren, *Heleodytes c. capistratus*
*Vol. III	p. 504: above chestnut
	Chiapas Cactus Wren, *Heleodytes chiapensis*
*Vol. III	p. 416: above chestnut
	Chestnut-backed Chickadee, *Penthistes r. rufescens*
*Vol. III	p. 80: forehead (broadly) chestnut
	American Barn Swallow, *Hirundo erythrogastra*
*Vol. III	p. 79: forehead (broadly) chestnut
	European Chimney Swallow, *Hirundo rustica*
*Vol. II	p. 751: crown chestnut
	Black-cheeked Warbler, *Basileuterus melanogneys*
*Vol. II	p. 743: crown-patch chestnut
	♂ Bell's Warbler, *Basileuterus b. belli*
Vol. II	p. 612: forehead, crown bright chestnut in spring and summer (grayish brown in winter)
	Palm Warbler, *Dendroica palmarum*
Vol. II	p. 592: pileum rich chestnut; chest and sides light chestnut (but breast is buffy, not bay)
	♂ Bay-breasted Warbler, *Dendroica castanea*
Vol. II	p. 589: broad side stripe rich chestnut
	♂ Chestnut-sided Warbler, *Dendroica pensylvanica*
Vol. II	p. 529: pileum varies rufous tawny to nearly chestnut
	♂ Bryant's Yellow Warbler, *Dendroica b. bryanti*
Vol. II	p. 275: below rich chestnut, deepening to bay (spring)
	♂ Orchard Oriole, *Icterus spurius*
*Vol. II	p. 73: nuchal band chestnut
	♂ Cozumel Spindalis, *Spindalis benedisti*
*Vol. II	p. 70: nuchal band chestnut
	♂ Black-backed Spindalis, *Spindalis z. zena*
Vol. II	p. 52: pileum ♂ dark chestnut or rich seal brown; ♀ more chestnut, sometimes bright chestnut
	Hooded Tanager, *Calospiza cucullata*
*Vol. I	p. 246: pileum chestnut
	Rufous-crowned Sparrow, *Amophila r. ruficeps*

Citations for Bay

See citations for Chestnut (Vol. VII, pp. 484, 56, 48).

Vol. VIII	p. 9: above bay or maroon (rump dark maroon)
	Central American Jacana, *Jacana s. spinosa*

Vol. VII	p. 316: above chestnut or bay
	Scaled Pigeon, *Lepidoenas speciosa*
Vol. VII	p. 138: forehead dark claret brown or bay (dark brownish red)
	Severe's Macaw, *Ara severa*
Vol. VII	p. 44: above general color between chestnut and bay (rich reddish brown)
	Panama Cuckoo, *Coccycua rutila panamensis*
Vol. II	p. 275: below rich chestnut, deepening to bay (in spring)
	♂ Orchard Oriole, *Icterus spurius*
Vol. II	p. 279: head, neck, chest rich dark chestnut or bay
	Martinique Oriole, *Icterus bonana*
Vol. II	p. 144: above dull brownish red, deep brick red, liver brown or bay
	♂ Mexican Ant Tanager, *Phoenicothraupis rubica rubicoides*
Vol. II	p. 43: head ♂ reddish chestnut or bay; ♀ duller chestnut
	Blue-rumped Green Tanager, *Calospiza gyroloides*

Citations for Auburn

Vol. X	p. 313: general color rich auburn to chestnut (♂ in ruddy phase, ♀ more chestnut)
	Eastern Bobwhite, *Colinus v. virginianus*
Vol. IX	p. 124: head, breast reddish auburn to chestnut
	Rufous-necked Wood-Rail, *Arimides axillaris*
Vol. IX	p. 115: occiput bright auburn
	Mexican Wood-Rail, *Arimides cajanea mexicana*

Citations for Brick Red

See citations for Bay (Vol. II, p. 144).

Vol. XI	p. 195: breast, thighs light brick red
	Mexican Sharp-shinned Hawk, *Accipiter striatus suttoni*
Vol. II	p. 149: above deep brick red
	♂ Tobasco Ant Tanager, *Phoenicothraupis salvini littoralis*
Vol. II	p. 148: above brick red or liver brown
	♂ Salvin's Ant Tanager, *Phoenicothraupis s. salvini*
Vol. II	p. 98: wings broadly edged brownish red, nearly brick red
	♂ Rose-throated Tanager, *Piranga r. roseo-gularis*
Vol. II	p. 86: above rich brownish red, between madder brown and burnt sienna
	♂ Brick-red Tanager, *Piranga t. testacea*

94

Cinnamon-Brown ~ Color 33

Cinnamon-Brown, a hyphenated color name, is shown in *Color Standards*. The hyphen is essential to distinguish the name from descriptive terminology. The name should not be confused with Cinnamon or Cinnamon-Rufous. Cinnamon-Brown is cited eleven times in Bulletin 50; a related name, Brussels Brown, only five times. The name is here grouped with Brussels Brown, Argus Brown, and Dresden Brown.

The COLOR GUIDE swatch chosen for Cinnamon-Brown has a 7.0YR 4.0/4.0 notation.

Notations of Colors Related to Cinnamon-Brown

Color Name	Ridgway Notation	Hamly Notation	Villalobos Notation
Cinnamon-Brown	XV 15'k	7.5 YR 3.8/4.0	OOS-7-6°
Cinnamon-Brown, measure of 1912		6.9 YR 4.0/3.4*	
Cinnamon-Brown, Guide swatch		7.0 YR 4.0/4.0*	
Argus Brown	III 13m	6.0 YR 3.8/4.0	OOS-6-8°
Brussels Brown	III 15m	7.5 YR 3.5/4.0	OOS-6-7°
Brussels Brown, measure of 1912		7.7 YR 3.7/4.1*	
Dresden Brown	XV 17'k	10.0 YR 3.8/4.0	OOS-7-5°

Citations for Cinnamon-Brown

Vol. X   p. 156: above cinnamon-brown to dark brussels (plus spots)
♂ Eastern Ruffed Grouse, *Bonasa u. umbellus* (brown phase)

Vol. VII   p. 424: back, scapulars cinnamon-brown, walnut, or rood's brown
Ruddy Ground Dove, *Chaemepelia r. rufipennis*

Vol. VI   p. 734: above general color snuff brown to cinnamon-brown
Strickland's Owl, *Lophostrix stricklandi* (pale phase)

Vol. V   p. 173: nape, back cinnamon-brown or between russet and raw umber
Mexican Xenops, *Xenops genibarbis mexicanus* and
Streaked Xenops, *Xenops rutilus heterurus*

Vol. IV   p. 823: above deep cinnamon-brown or russet (specimen 494223 measured 8.0YR 2.7/3.2)
Mexican Lathria, *Lathria u. unirufa*

Vol. IV   p. 820: above cinnamon-brown or tawny russet (specimen
             390756 measured 7.7YR 2.9/3.7)
             Rufous Lipaugus, *Lipaugus h. holerythrus*
Vol. IV   p. 64: above tawny, from near cinnamon-brown to more isabel-
             line (in spring, fall more tawny)
             Wilson's Thrush, *Hylocichla f. fuscens*
*Vol. IV   p. 37: back cinnamon-brown (in spring, brighter in fall)
             Wood Thrush, *Hylocichla mustelina*
Vol. IV   p. 29: above russet, varying tawny olive or raw umber to cin-
             namon-brown; tail cinnamon-brown to russet
             Nightingale Thrush, *Catharus m. melpomene*
Vol. IV   p. 39: tail dull cinnamon-brown or cinnamomeous wood brown
             Kadiak Dwarf Thrush, *Hylocichla g. guttata*
Vol. III   p. 505: rump dull cinnamon-rufous or cinnamon-brown (plus
             streaks)
             Hooded Cactus Wren, *Heleodytes c. capistratus*

Citations for Brussels Brown

See citations for Cinnamon-Brown (Vol. X, p. 56).

Vol. XI   p. 744: rectrices argus brown to brussels brown (not hazel)
             (rufous phase)
             ♂ Cuban Sparrow Hawk, *Falco sparverius sparverioides*
Vol. XI   p. 238: rectrices bright hazel to brussels brown
             Eastern Red-tailed Hawk, *Buteo jamaicensis borealis*
*Vol. X   p. 212: forehead brussels brown
             Greater Prairie Hen, *Tympanuchus cupido pinnatus*
Vol. IX   p. 108: head, neck brussels brown to rufescent sudan brown
             Jamaican Uniform Crake, *Amaurolimnas c. concolor*

Russet ~ Color 34

Russet is grouped with Cameo Brown, Walnut Brown, and Rood's Brown. All four colors are closely correlated by Hamly and Villalobos. Two other colors are listed, but their similarity is not quite sufficient to permit their inclusion in the Russet "family." For example, the Villalobos notation of SO-9-3° for Army Brown is hardly distinguishable from his SO-8-4° for Russet, whereas the Hamly notations are more clearly separable. Russet is related to Cinnamon-Brown, and it may be difficult to separate one from the other.

Nearly thirty citations from Bulletin 50 are listed. They link Russet with darker colors, such as Raw Umber, more than with such bright colors as Cinnamon-Rufous and Tawny.

The COLOR GUIDE swatch chosen for Russet has a 5.0YR 4.0/4.0 notation.

Notations of Colors Related to Russet

Color Name	Ridgway Notation	Hamly Notation	Villalobos Notation
Cameo Brown	XXVIII 7''k	2.5 YR 4.0/3.0	SSO-7-4°
Walnut Brown	XXVIII 9''k	2.5 YR 4.0/3.0	SO-8-4°
Rood's Brown	XXVIII 11''k	4.0 YR 4.5/3.0	SO-8-4°
Russet	XV 13'k	5.0 YR 3.8/4.0	SO-8-4°
Russet, measure of 1912		3.9 YR 4.2/4.5*	
Russet, Guide swatch		5.0 YR 4.0/4.0*	
Russet-Vinaceous	XXXIX 9'''	10.0 R 5.0/3.0	SSO-9-5°
Army Brown	XL 13'''i	5.0 YR 4.4/2.5	SO-9-3°

Citations for Russet

Vol. XI	p. 163: thighs between russet and chestnut Bicolored Hawk, *Accipiter b. bicolor*
*Vol. XI	p. 132: thighs bright russet, plus barring ♀South American Snail Kite, *Rostrhamus s. sociabilis* and ♀Everglade Snail Kite, *R. s. plumbeus*
*Vol. XI	p. 117: breast nearly solid russet (♀ more so) Banded Double-toothed Hawk, *Harpagus bidentatus fasciatus*
*Vol. X	p. 46: under wing and under tail coverts russet Rufous-tailed Chachalaca, *Ortalis ruficauda*
Vol. X	p. 49: belly more hazel than russet Northern Rufous-bellied Chachalaca, *Ortalis wagleri griseiceps*
Vol. X	p. 48: belly deep hazel to russet Rufous-bellied Chachalaca, *Ortalis w. wagleri*

Vol. VIII	p. 303: pileum dull cinnamon-rufous or hazel, nearly mikado brown or light russet (in summer)
	♂ Spoon-billed Sandpiper, *Eurynorhynchus pygmeus*
Vol. VIII	p. 192: below light russet or mikado brown (in summer)
	Hudsonian Godwit, *Vetola haemastica*
Vol. VII	p. 225: above and below deep orange brown, nearly russet
	St. Vincent Parrot, *Amazona guildingii*
Vol. VII	p. 44: throat, chest between tawny and russet
	Panama Cuckoo, *Coccycua rutila panamensis*
Vol. V	p. 776: rump and tail coverts nearly tawny olive or russet
	♀ Jalapa Trogon, *Trogonurus puella*
Vol. V	p. 291: above between raw umber and mummy to nearly russet
	Brown Dendrocincla, *Dendrocincla lafresnayii ridgwayi*
*Vol. V	p. 280: back russet, sometimes tinged with olive
	Mexican Sittasomus, *Sittasomus s. sylvioides*
Vol. V	p. 275: back raw umber or russet
	Northern Wedgebill, *Glyphorynchus cuneatus pectoralis*
Vol. V	p. 271: back more fulvous or russet, between raw umber and russet or mars brown
	Venezuelan Sickle Bill, *Campylorhamphus venezuelensis*
Vol. V	p. 229: pileum, nape dull cinnamon-rufous or russet, plus bars
	Barred Woodhewer, *Dendrocolaptes s. sancti-thomae*
Vol. V	p. 175: nape, back between russet and raw umber (= dull cinnamon-brown)
	Streaked Xenops, *Xenops rutilus heterurus*
Vol. V	p. 175: nape, back cinnamon-brown or between russet and raw umber
	Mexican Xenops, *Xenops genibarbis mexicanus*
Vol. V	p. 56: pileum, nape russet or chestnut-brown
	♀ Northern Antvireo, *Dysithamnus mentalis septentrionalis*
Vol. IV	p. 823: above deep cinnamon-brown or russet
	Mexican Lathria, *Lathria u. unirufa*
*Vol. IV	p. 765: tail russet
	♂ Rufous Manakin, *Laniocera rufescens*
Vol. IV	p. 52: above between raw umber and russet (in autumn)
	Russet-backed Thrush, *Hylocichla u. ustulata*
Vol. IV	p. 37: pileum tawny or russet (in summer)
	Wood Thrush, *Hylocichla mustelina*
Vol. IV	p. 29: above russet, but varies between tawny olive and raw umber to cinnamon-brown; tail cinnamon-brown to russet
	Nightingale Thrush, *Catharus m. melpomene*
Vol. IV	p. 28: pileum, nape deep russet, or from nearly mars brown to almost raw umber
	Frantzius' Nightingale Thrush, *Catharus f. frantzii*
Vol. IV	p. 26: pileum, nape deep russet, or between russet and mars brown
	Russet Nightingale Thrush, *Catharus o. occidentalis*
*Vol. III	p. 602: above tawny, between dark broccoli brown and russet; tail russet
	Unalaska Wren, *Olbiorchilus alascensis*

*Vol. III   p. 590: above tawny, between mummy and russet; pileum deep russet

Irazú Wren, *Troglodytes ochraceus*

Vol. II   p. 380: below cinnamon-rufous or russet

♂ Mexican Diglossa, *Diglossa baritula*

Hazel ~ Color 35

Hazel occurs in *Color Standards* close to Russet and Cinnamon-Brown, but has a ruddier effect. Although Hazel is not used in Bulletin 50 as often as might be expected, there are a fair number of citations. These are supplemented by a few citations for Mikado Brown, a very similar color. Other colors here merged within the Hazel group are rarely used in Bulletin 50, and no acceptable citations have been found for them.

The COLOR GUIDE swatch chosen for Hazel has a 2.5YR 4.5/5.5 notation.

Notations of Colors Related to Hazel

Color Name	Ridgway Notation	Hamly Notation	Villalobos Notation
Vinaceous-Russet	XXVIII 7″i	2.5 YR 4.5/4.0	SO-10-4°
Cacao Brown	XXVIII 9″i	2.5 YR 4.6/4.0	SO-10-4°
Hazel	XIV 11′k	2.5 YR 4.4/5.5	OOS- 8-8°
Hazel, measure of 1912		4.8 YR 4.4/5.7*	
Hazel, measure of 1886		2.0 YR 5.1/6.5*	
Hazel, Guide swatch		2.5 YR 4.5/5.5*	
Kaiser Brown	XIV 9′k	2.5 YR 3.8/6.0	SO- 8-5°
Pecan Brown	XXVIII 11″i	5.0 YR 5.0/4.5	SO-10-5°
Mikado Brown	XXIX 13″i	5.0 YR 5.0/5.0	OOS- 9-6°
Mikado Brown, measure of 1912		4.8 YR 5.1/5.1*	

Citations for Hazel

Vol. XI   p. 723: above pale cinnamon-rufous to deep hazel, plus barring; head patch mikado brown to cinnamon-rufous or hazel
♂ Northern Sparrow Hawk, *Falco s. sparverius*

Vol. XI   p. 675: lower belly, thighs deep hazel to deep cinnamon-rufous
Bat Falcon, *Falco a. albigularis*

Vol. XI   p. 671: lower belly, thighs called chestnut (as generalization), then described as deep, bright hazel (breast paler)
Orange-breasted Falcon, *Falco deiroleucus*

Vol. XI   p. 246: rectrices orange cinnamon to dark hazel
Western Red-tailed Hawk, *Buteo jamaicensis calurus*

Vol. XI   p. 238: rectrices bright hazel to brussels brown
Eastern Red-tailed Hawk, *B. j. borealis*

*Vol. X   p. 341: belly hazel
Salvin's Bobwhite, *Colinus virginianus salvini*

*Vol. X	p. 338: below orange-cinnamon to hazel; belly uniformly hazel Guatemalan Bobwhite, *C. v. insignis*
Vol. X	p. 49: belly more hazel than russet Northern Rufous-bellied Chachalaca, *Ortalis wagleri griseiceps*
Vol. X	p. 48: belly deep hazel to russet Rufous-bellied Chachalaca, *Ortalis w. wagleri*
Vol. IX	p. 108: below rich tawny to hazel Jamaican Uniform Crake, *Amaurolimnas c. concolor*
Vol. VIII	p. 303: pileum dull cinnamon-rufous or hazel ♂ Spoon-billed Sandpiper, *Eurynorhynchus pygmeus*
Vol. VIII	p. 250: head and underparts chestnut rufous or hazel (in summer) ♂ Curlew Sandpiper, *Erolia ferruginea*
Vol. VII	p. 67: pileum, crest dull cinnamon-rufous or hazel Northern Striped Cuckoo, *Tapera naevia excellens*
Vol. VI	p. 784: above deep cinnamon-rufous or hazel (rufous phase) Jardine's Pygmy Owl, *Glaucidium jardinii*
Vol. IV	p. 765: dull cinnamon-rufous or hazel Rufous Manakin, *Laniocera rufescens*

Citations for Mikado Brown

See citations for Hazel (Vol. XI, p. 723).

Vol. XI	p. 444: face, nape side-neck, upper scapulars snuff to mikado brown American Crested Eagle-Hawk, *Spizaetus ornatus vicarius*
Vol. VIII	p. 303: pileum dull cinnamon-rufous or hazel, nearly mikado brown or light russet (in summer) ♂ Spoon-billed Sandpiper, *Eurynorhynchus pygmeus*
Vol. VIII	p. 232: below buffy cinnamon, vinaceous-cinnamon, or mikado (in summer) ♂ Knot, *Canutus canutus*
Vol. VIII	p. 192: below light russet or mikado brown (summer), plus barring Hudsonian Godwit, *Vetola haemastica*
Vol. VII	p. 352: below bright mikado brown or deep orange-cinnamon ♂ Socorro Mourning Dove, *Zenaidura graysoni*

Amber ~ Color 36

Amber, as a physical material, is a fossilized resin. Amber resins occur in a wide variety of colors, varying from pale yellow to dark golden brown. Ridgway's Amber Brown, here called simply Amber, is about halfway between the swatches for Sanford's Brown and Sudan Brown. It is clearly close enough to the former to warrant placing them in the same family, but not close enough to call them equivalents.

Ridgway rarely uses Amber in Bulletin 50, and no acceptable citations were found for it. Neither is it greatly used by other ornithologists, although Phillips (1962) found it useful in his paper on thrushes of the genus *Catharus,* where he relied on the colors depicted in *Color Standards.* The following colors are used in the paper; those also shown in the NATURALIST'S COLOR GUIDE are in boldface; those marked with an asterisk are mentioned but not depicted:

Amber Brown	Mars Brown*	**Drab**
Antique Brown	**Olive-Brown**	Isabella Color*
Argus Brown*	Prout's Brown*	Mouse Gray*
Brussels Brown*	Sudan Brown*	**Russet**
Buffy Brown*	Wood Brown*	**Raw Umber**
Cinnamon-Brown	Auburn*	Saccardo's Umber
Dresden Brown*	**Brownish-Olive**	Tawny-Olive*
Hair Brown*	Cream-Buff*	

Despite its infrequent use, Amber helps to fill a color gap between Hazel and Ridgway's Cinnamon-Rufous.

The COLOR GUIDE swatch chosen for Amber has a 5.0YR 4.0/8.0 notation.

Notations of Colors Related to Amber

Color Name	Ridgway Notation	Hamly Notation	Villalobos Notation
Sanford's Brown	II 11k	2.5 YR 4.2/8.0	SO-8-11°
Amber Brown	III 13k	5.0 YR 3.8/8.0	OOS-8-10°
Amber Brown, measure of 1912		5.8 YR 4.1/6.0*	
Amber, Guide swatch		5.0 YR 4.0/8.0*	

Citations for Amber

No useful citations for Amber were found in Bulletin 50.

Antique Brown ~ Color 37

Antique Brown and Sudan Brown are here grouped together. Their notations and visual appearance are closely alike. Neither of the names is especially descriptive, but a choice must be made and Antique Brown seems better than Sudan Brown.

Neither color is used greatly by ornithologists, although they both appear in Phillip's 1962 paper. In common with Amber, they fill a gap between Hazel and Cinnamon-Rufous.

Antique Brown is used because a substantial number of bird specimens tested by spectrophotometer measured close to the color, although they are described in Bulletin 50 under different colors. In three instances they are called cinnamon-rufous, in others chestnut-rufous, deep vinaceous cinnamon, bright chestnut, and cinnamon suffused with chestnut. A more frequent use of Antique Brown is indicated.

The COLOR GUIDE swatch chosen for Antique Brown has a 7.5YR 4.5/5.0 notation.

Notations of Colors Related to Antique Brown

Color Name	Ridgway Notation	Hamly Notation	Villalobos Notation
Sudan Brown	III 15k	7.5 YR 4.5/6.0	O-8-12°
Antique Brown	III 17k	7.5 YR 4.4/5.0	O-8- 9°
Antique Brown, measure of 1912		8.4 YR 4.3/5.4*	
Antique Brown, Guide swatch		7.5 YR 4.5/5.0*	

Citations for Antique Brown

No useful citations for Antique Brown were found in Bulletin 50.

Tawny ~ Color 38

Tawny is grouped with Ochraceous-Tawny and Orange-Cinnamon. It is frequently used in Bulletin 50, where some of the citations are confusing, especially when Tawny is called "nearly raw umber" or "between mummy brown and russet," all of which are surely much darker. Included in the citations are two for tawny-cinnamon, whicn is not shown in *Color Standards*. The tendency to use such terms increases in Bulletin 50 among the lighter colors. Citations for tawny-ochraceous are included with those for ochraceous-tawny on the assumption that they are equivalent.

Palmer (1962) shows a swatch for Tawny. Measured by spectrophotometer, it is a good representation of the color.

The Color Guide swatch chosen for Tawny has a 5.0YR 5.0/7.0 notation.

Notations of Colors Related to Tawny

Color Name	Ridgway Notation	Hamly Notation	Villalobos Notation
Tawny	XV 13′i	5.0 YR 5.0/6.5	OOS-10-7°
Tawny, measure of 1912		5.0 YR 5.1/6.5*	
Tawny, measure of 1886		5.1 YR 5.7/8.4*	
Ochraceous-Tawny	XV 15′i	8.5 YR 5.5/7.0	O-12-8°
Ochraceous-Tawny, measure of 1912		6.3 YR 5.2/6.2*	
Tawny-Ochraceous, measure of 1886		7.0 YR 6.1/8.2*	
Tawny, Palmer		5.0 YR 5.1/6.9*	OOS-10-7°
Tawny, Guide swatch		5.0 YR 5.0/7.0*	
Orange-Cinnamon	XXIX 13″	5.0 YR 5.8/7.0	OOS-11-7°

Citations for Tawny

Vol. IX	p. 108: below rich tawny to hazel
	Jamaican Uniform Crake, *Amaurolimnas c. concolor*
Vol. VII	p. 84: below light tawny or buffy brown
	Salvin's Ground Cuckoo, *Neomorphus salvini*
Vol. VII	p. 62: belly tawny, ochraceous-tawny, or deep ochraceous
	Porto Rican Lizard-Cuckoo, *Saurothera vieilloti*
Vol. VII	p. 44: throat, chest between tawny and russet
	Panama Cuckoo, *Coccycua rutila panamensis*
Vol. VI	p. 468: belly rufous tawny
	Greater Rufous Motmot, *Urospatha martii semirufa*

Vol. VI	p. 366: below rufous tawny Black-chinned Jacamar, *Galbula melanogenia*
Vol. VI	p. 363: ♂ below tawny or rufous tawny; ♀ throat-patch below tawny or tawny-ochraceous Great Jacamar, *Jacamerops aurea*
Vol. VI	p. 141: nape, crest tawny to clay color, or dull ochraceous Chestnut-colored Woodpecker, *Celeus castaneus*
Vol. V	p. 214: back, rump tawny or mummy brown Vera Paz Automolus, *Automolus v. veraepacis*
Vol. V	p. 203: back tawny, nearly raw umber Ochraceous Philydor, *Philydor penerythus*
*Vol. V	p. 62: pileum bright tawny ♀Surinam Antwren, *Myrmotherula surinamensis*
Vol. V	p. 22: above tawny, near raw umber or tawny-olive Tawny Antshrike, *Thamnistes a. anabatinus*
Vol. IV	p. 840: above rufous tawny; below buff or tawny-buff Cinnamon Becard, *Pachyrhamphus cinnamomeus*
Vol. IV	p. 839: above bright tawny or rufous tawny ♀Cinereus Becard, *Pachyrhamphus cinereus*
Vol. IV	p. 64: above tawny; in spring from near cinnamon-brown to more isabelline; in autumn brighter, more tawny or cinnamomeous Wilson's Thrush, *Hylocichla f. fuscescens*
Vol. IV	p. 67: above less tawny than *H.f.f.*, deep isabella to near broccoli Willow Thrush, *Hylocichla fuscescens salicicola*
Vol. III	p. 602: above tawny, between dark broccoli and russet Unalaska Wren, *Olbiorchilus alascensis*
Vol. III	p. 590: above tawny between mummy brown and russet Irazú Wren, *Troglodytes ochraceus*
Vol. II	p. 529: pileum rufous tawny to nearly chestnut ♂Bryant's Yellow Warbler, *Dendroica b. bryanti*
Vol. II	p. 136: bushy crest bright tawny or orange tawny ♂Tawny-crested Tanager, *Tachyphonus delatri*
*Vol. II	p. 29: median underparts deep tawny ♂Gould's Euphonia, *Euphonia gouldi*
Vol. II	p. 20: belly light tawny or deep ochraceous ♀Cabanis Euphonia, *Euphonia gracilis*
*Vol. II	p. 17: pileum rich tawny ♂Tawny-capped Euphonia, *Euphonia anneae*

Citations for Orange-Cinnamon

Vol. XI	p. 246: rectrices orange-cinnamon to dark hazel Western Red-tailed Hawk, *Buteo jamaicensis calurus*
Vol. XI	p. 226: above bright orange-cinnamon to ferruginous (light phase) Ferruginous Rough-legged Hawk, *Buteo regalis*
*Vol. XI	p. 205: pileum, nape, thighs orange-cinnamon (plus streaks and bars) Rufous-headed Hawk, *Heterospizias m. meridionalis*

Vol. X	p. 338: below orange-cinnamon to hazel, belly uniform hazel
	♂ Guatemalan Bobwhite, *Colinus virginianus insignis*
Vol. VIII	p. 291: face, throat light cinnamon-rufous or orange-cinnamon
	♂ Rufous-necked Sandpiper, *Pisobia ruficollis*
Vol. VII	p. 352: below bright mikado brown or deep orange-cinnamon
	♂ Socorro Mourning Dove, *Zenaidura graysoni*

Citations for Ochraceous-Tawny

See citations for Tawny (Vol. VII, p. 62; Vol. VI, p. 363).

Vol. XI	p. 671: chest, breast bright ochraceous-tawny to pale cinnamon (but called chestnut or rufous by others)
	Orange-breasted Falcon, *Falco deiroleucus*
Vol. VII	p. 71: below tawny-ochraceous to clay or cinnamon-ochraceous
	Rufous-rumped Cuckoo, *Morococcyx e. erythopygus*
*Vol. VII	p. 56: belly, thighs tawny-ochraceous (light)
	Haitian Hyetornis, *Hyetornis rufigularis*
*Vol. V	p. 354: tail tawny-ochraceous
	♂ Mexican Royal Flycatcher, *Onychorhynchus m. mexicanus*
*Vol. V	p. 293: below dull tawny-ochraceous; chest darker, deep tawny
	Ruddy Dendrocincla, *Dendrocincla h. homochroa*
Vol. V	p. 214: chest, throat deep tawny-ochraceous
	Vera Paz Automolus, *Automolus v. veraepacis*
Vol. V	p. 205: below dull tawny-ochraceous to raw sienna
	Dusky-winged Philydor, *Philydor fuscipennis*
Vol. IV	p. 820: below dull tawny-ochraceous to dull tawny
	Rufous Lipaugus, *Lipaugus h. holerythrus*
Vol. IV	p. 97: below pale cinnamon-rufous, some tawny-ochraceous
	♀ American Robin, *Planesticus m. migratorius*
Vol. III	p. 533: below tawny-ochraceous, deepest on chest, paler on belly; some are tawny-buff
	Tawny-bellied Wren, *Pheugopedius hyperythrus*

Citations for Tawny-Cinnamon

Tawny-cinnamon is not shown in *Color Standards.*

Vol. X	p. 345: below light tawny-cinnamon
	♂ Masked Bobwhite, *Colinus virginianus ridgwayi*
Vol. IX	p. 115: breast, upper belly tawny-cinnamon to pale buckthorn
	Mexican Wood-rail, *Aramides c. cajanea*

Cinnamon ~ Color 39

Cinnamon is shown by Ridgway in both color guides, and there are about a dozen citations for it in Bulletin 50. It has an individual quality that matches quite closely to commercial stick cinnamon.

Palmer's swatch for Cinnamon has the same notation as that given by Villalobos for Ridgway's swatch: OOS-13-6°. It does not differ greatly from the somewhat more bland choice of color in this guide.

The COLOR GUIDE swatch chosen for Cinnamon has a 7.5YR 6.0/6.0 notation.

Notations of Colors Related to Cinnamon

Color Name	Ridgway Notation	Hamly Notation	Villalobos Notation
Cinnamon	XXIX 15″	7.5 YR 6.2/6.0	OOS-13-6°
Cinnamon, measure of 1912		7.0 YR 5.8/5.7*	
Cinnamon, measure of 1886		7.9 YR 5.8/5.8*	
Cinnamon, Palmer glossy		6.2 YR 5.6/5.4*	OOS-13-6°
Cinnamon, Guide swatch		7.5 YR 6.0/6.0*	

Citations for Cinnamon

Vol. XI   p. 724: below light pinkish cinnamon to cinnamon
♂ Northern Sparrow Hawk, *Falco s. sparverius*

Vol. XI   p. 671: lower throat, breast bright ochraceous-tawny to pale cinnamon; lower belly, thighs deep bright hazel (but I believe the description is in error and that the colors of the Orange-breasted Falcon are similar to those of the Bat Falcon, which were called chestnut by Blake)
Orange-breasted Falcon, *Falco deiroleucus*

Vol. IX   p. 169: below cinnamon, suffused with deep chestnut
Ruddy Rail, *Laterallus r. ruber*

*Vol. VIII   p. 437: head, neck, chest light cinnamon (in summer)
American Avocet, *Recurvirostra americana*

Vol. VIII   p. 232: below buffy cinnamon, vinaceous-cinnamon, or mikado brown (in summer)
♂ Knot, *Canutus canutus*

Vol. VIII   p. 184: in general cinnamon, or between pinkish buff and vinaceous-buff
Marbled Godwit, *Vetola fedoa*

Vol. VIII   p. 113: pileum, throat band sayal brown, or dull cinnamon
Cinnamomeous Plover, *Pagolla wilsonia cinnamomina*

*Vol. VI   p. 382: throat cinnamon, near russet centrally
Russet-throated Puffbird, *Hypnelus r. ruficollis*

Cinnamon-Rufous ~ Color 40

Cinnamon-Rufous, a hyphenated color name, is shown in both *Color Standards* and *Nomenclature*. It is used very frequently in Bulletin 50, and more than sixty citations are listed. Ten of the birds cited were measured by spectrophotometer to determine how closely they matched the notation chosen for Cinnamon-Rufous—2.5YR 5.0/8.0. In no instance did they come near it; they measured closer to other Ridgway colors, such as Raw Umber, Cinnamon-Brown, Russet, Antique Brown, and Hazel. They ranged from 5.0YR to 7.5YR in hue (not 2.5YR) and from /3.0 to /5.6 in chroma (not /8.0). Individual measurements are included in the citations.

It is probable that other ornithologists have also erroneously used Cinnamon-Rufous, perhaps following Ridgway's descriptions uncritically rather than making their own comparisons with color guides. One such possibility is present in a series of papers on flycatchers of the genus *Myiarchus* (Lanyon, 1960, 1961; Phillips, 1966, 1970). The inner webs of some of the rectrices of these birds are of a conspicuous color that flashes brightly when the tail is spread. The distinctive patterns of these areas are used to distinguish certain *Myiarchus* species from each other. The authors call the areas Cinnamon-Rufous (Color 40), in agreement with Ridgway, but visual and spectrophotometric measurements showed that the color is close to Antique Brown (Color 37); it is a less flashy color than Cinnamon-Rufous. Perhaps the error derives from the brighter appearance of these areas when seen flashing in direct sunlight.

A few measurements were made on birds with plumage described as chestnut and cinnamon-brown, which resulted in similar discrepancies. These discrepancies permit one to generalize the following:

1. At least some descriptions given by Ridgway are less accurate than they are presumed to be.

2. The application of a color name to a plumage creates a sense of accuracy that is not warranted.

3. Some type of scientifically controlled measurement is essential when accuracy is intended.

A different type of study entailed the examination of descriptions by other ornithologists compared with Ridgway's usage of Cinnamon-Rufous and other rufescent colors (see table on next page).

An additional comparison is suggested by a recent paper (Willis, 1972), which states that "as in many woodcreepers the flight and tail feathers [of the Plain-brown Woodhewer] are rufous." Ridgway

Common Name	Ridgway	Blake	Land	Peterson
Ruddy Crake	bright chestnut	bright chestnut	rufous	deep rufous
Cinnamon Hummingbird (below)	light cinnamon-rufous or deep vinaceous cinnamon	cinnamon-rufous	cinnamon	cinnamon
(tail)	deep cinnamon-rufous	deep cinnamon-rufous or chestnut	rufous	rufous
Fawn-breasted Hummingbird	cinnamon-rufous or vinaceous-cinnamon	cinnamon-buff	tawny	buff to tawny
Rufous-tailed Hummingbird	chestnut below	deep chestnut	rufous	—
Rufous Hummingbird	cinnamon-rufous	rufous	—	rufous
Allen's Hummingbird	deep cinnamon-rufous	rufous	—	rufous
Ringed Kingfisher	deep cinnamon-rufous	rufous	rufous	rufous
Belted Kingfisher	cinnamon-rufous	rufous	rufous	—
Amazon Kingfisher	deep cinnamon-rufous or rufous chestnut	rufous	rufous	rufous
Green Kingfisher	rufous chestnut or chestnut rufous	rufous	chestnut	rufous
Barred Antshrike	♀ cinnamon-rufous or tawny chestnut	cinnamon-rufous	rufous	bright cinnamon above, tawny below
Black-faced Antthrush	chestnut	rufous	rufous	rufous
Rufous Mourner	cinnamon-brown or tawny russet	cinnamon-brown	rufous	rufous
Rufous Piha	deep cinnamon-brown or russet	cinnamon-brown	rufous	rufous
Cinnamon Becard	rufous tawny	cinnamon or rufous	rufous	rufous
Crested Flycatcher	cinnamon-rufous	cinnamon-rufous	rufous	rusty-tailed
American Robin	♂ deep cinnamon rufous	cinnamon-rufous	rufous	brick red
Rufous-sided Towhee	cinnamon-rufous or rufous chestnut	cinnamon-rufous	rufous	rufous

describes the species as "cinnamon-rufous or rufous chestnut." Measurement of a specimen showed a 4.8YR 2.7/4.0 notation, which places the color between Russet and Chestnut—not Rufous or Cinnamon-Rufous.

Rufous is represented by a swatch in *Color Standards,* and Palmer (1962) shows a nearly identical swatch. They have approximately 10.0R 5.5/10.0 notations. The swatches depict a very vivid reddish orange color. In *Color Standards* Rufous is adjacent to Carnelian Red, Apricot Orange, and Ferruginous. It is hardly ever used in Bulletin 50; only one citation was found (Vol. V, p. 416, Forrer's Hummingbird) where it was used as a specific color. Ridgway usually uses it in a generalized sense, possibly meaning rufescence instead of a color.

The facing tabulation shows a strong tendency to call the plumages simply rufous, but all of the comparisons with color swatches are for a much darker color than Ridgway's Rufous, ranging from Cinnamon-Rufous through Russet to nearly Chestnut. Some vaguely reddish brown or rusty brown quality is evidently being called rufous, in which case the term *rufescence* would be preferred. But even rufescence implies some fairly definite color, and because Ridgway's Rufous swatch is far too vivid, his Cinnamon-Rufous swatch is used for that purpose in the NATURALIST'S COLOR GUIDE.

The COLOR GUIDE swatch chosen for Cinnamon-Rufous has a 2.5YR 5.0/8.0 notation.

Notations of Colors Related to Cinnamon-Rufous

Color Name	Ridgway Notation	Hamly Notation	Villalobos Notation
Cinnamon-Rufous	XIV 11'i	2.5 YR 5.0/8.0	OOS-10-8°
Cinnamon-Rufous, measure of 1912		3.5 YR 4.9/6.6*	
Cinnamon-Rufous, measure of 1886		3.0 YR 5.1/7.2*	
Cinnamon-Rufous, Guide swatch		2.5 YR 5.0/8.0*	

Citations for Cinnamon-Rufous

Vol. XI   p. 723: head-patch mikado brown to cinnamon-rufous or hazel, upper parts pale cinnamon-rufous to deep hazel
♂ Northern Sparrow Hawk, *Falco s. sparverius*

Vol. XI   p. 675: lower belly, thighs deep hazel to deep cinnamon-rufous (but called chestnut or rufous by others)
Bat Falcon, *Falco a. albigularis*

*Vol. VIII	p. 424: neck, chest cinnamon-rufous
	♀Northern Phalarope, *Lobites lobatus*
Vol. VIII	p. 303: pileum dull cinnamon-rufous or hazel, near mikado or light russet
	♂Spoon-billed Sandpiper, *Eurynorhynchus pygmeus*
Vol. VIII	p. 291: face, throat light cinnamon-rufous or orange-cinnamon
	♂Rufous-necked Sandpiper, *Pisobia ruficollis*
Vol. VII	p. 67: dull cinnamon-rufous or hazel
	Northern Striped Cuckoo, *Tapera naevia excellens*
Vol. VII	p. 55: below deep cinnamon-rufous or rufous chestnut
	Jamaican Hyetornis, *Hyetornis pluvialis*
*Vol. VII	p. 51: above light cinnamon-rufous (pileum paler, tail deeper)
	Mexican Squirrel-Cuckoo, *Piaya mexicana*
Vol. VI	p. 688: general color cinnamon-rufous or chestnut rufous (in rufescent phase)
	Florida Screech Owl, *Otus a. asio*
*Vol. VI	p. 478: belly cinnamon-rufous
	Turquoise-browed Motmot, *Eumota s. superciliosa*
*Vol. VI	p. 472: pileum, nape, face dull cinnamon-rufous
	Lesser Broad-billed Motmot, *Electron platyrhynchus minor*
*Vol. VI	p. 468: pileum, nape bright cinnamon-rufous; belly cinnamon-rufous or deep rufous tawny
	Greater Rufous Motmot, *Urospatha martii semirufa*
Vol. VI	p. 429: chest band deep rufous chestnut or chestnut rufous (specimen 43958 measured 5.0YR 3.4/5.6)
	♂Green Kingfisher, *Chloroceryle americana septentrionalis*
Vol. VI	p. 424: chest band cinnamon-rufous or rufous chestnut (specimen 769898 measured 5.5YR 4.0/5.3)
	♂Amazon Kingfisher, *Chloroceryle amazona*
Vol. VI	p. 415: breast band cinnamon-rufous
	♀Belted Kingfisher, *Ceryle a. alcyon*
Vol. VI	p. 409: below deep cinnamon-rufous (specimen 106215 measured 5.0YR 3.5/5.1)
	Ringed Kingfisher, *Streptoceryle t. torquata*
Vol. VI	p. 342: nuchal collar deep cinnamon-rufous or chestnut
	Collared Araçari, *Pteroglossus t. torquatus*
Vol. VI	p. 143: above cinnamon-rufous or rufous chestnut, plus bars
	Fraser's Woodpecker, *Celeus l. loricatus*
Vol. VI	p. 37: pileum, nape deep cinnamon-rufous or rufous chestnut
	Guatemalan Flicker, *Colaptes mexicanoides*
*Vol. V	p. 612: above cinnamon-rufous, plus glosses
	♂Rufous Hummingbird, *Selasphorus rufus*
Vol. V	p. 609: tail and coverts, facial area deep cinnamon-rufous or rufous chestnut
	♂Allen's Hummingbird, *Selasphorus alleni*
Vol. V	p. 419: below deeper cinnamon-rufous than next citation (specimen 394038 measured 6.6YR 4.0/4.5)
	Coral-billed Hummingbird, *Amizilis rutila corallirostris*
Vol. V	p. 416: below dull light cinnamon-rufous or deep vinaceous-cinnamon; tail deep cinnamon-rufous or rufous-chestnut (specimen 480386 measured 7.5YR 4.5/4.6 below and 5.1YR 3.0/4.0 at tail)

112

Cinnamomeous Hummingbird, *Amizilis r. rutila*

Vol. V   p. 414: below paler cinnamon-rufous, nearly vinaceous-cinnamon

Fawn-breasted Hummingbird, *Amizilis yucatanensis cerviniventris*

*Vol. V   p. 413: below cinnamon-rufous

Yucatan Hummingbird, *Amizilis y. yucatanensis*

Vol. V   p. 291: tail deep cinnamon-rufous or rufous chestnut (specimen 284105 measured 4.5YR 2.5/4.0)

Brown Dendrocincla, *Dendrocincla lafresnayei ridgwayi*

Vol. V   p. 288: tail deep cinnamon-rufous or rufous chestnut

Northern Dendrocincla, *Dendrocincla a. anabatina*

*Vol. V   p. 280: rump, tail deep cinnamon-rufous

Mexican Sittasomus, *Sittasomus s. sylvioides*

*Vol. V   p. 272: rump cinnamon-rufous

Costa Rican Sickle Bill, *Campylorhamphus borealis*

Vol. V   p. 264: rump, tail cinnamon-rufous or rufous chestnut

Streak-headed Woodhewer, *Picolaptes l. lineaticeps*

Vol. V   p. 259: rump, tail deep cinnamon-rufous or rufous chestnut

White-striped Woodhewer, *Picolaptes leucogaster*

Vol. V   p. 249: tail deep cinnamon-rufous or rufous chestnut

Stripe-throated Woodhewer, *Xiphorhynchus striatigularis*

Vol. V   p. 244: tail deep cinnamon-rufous or rufous chestnut

Swainson's Woodhewer, *Xiphorhynchus flavigaster*

Vol. V   p. 229: pileum, nape dull cinnamon-rufous or russet, plus bars

Barred Woodhewer, *Dendrocolaptes s. sancti-thomae*

*Vol. V   p. 194: pileum, nape cinnamon-rufous

Costa Rican Gray-breasted Synallaxis, *Synallaxis albescens latitabunda*

Vol. V   p. 189: chest deep cinnamon-rufous or chestnut rufous

Rufous-breasted Synallaxis, *Synallaxis erythrothorax*

*Vol. V   p. 185: head, face bright cinnamon-rufous

Lawrence's Spinetail, *Acrorchilus erythrops rufigenis*

Vol. V   p. 178: above cinnamon-rufous or rufous chestnut

Costa Rican Margarornis, *Margarornis rubiginosa*

*Vol. V   p. 175: rump, tail deep cinnamon-rufous

Streaked Xenops, *Xenops rutilus heterurus*

Vol. V   p. 45: above bright cinnamon-rufous or rufous chestnut

♀Barred-crested Antshrike, *Thamnophilus multistriatus*

Vol. V   p. 40: above cinnamon-rufous or tawny chestnut

♀Mexican Antshrike, *Thamnophilus doliatus mexicanus*

Vol. V   p. 37: tail cinnamon-rufous or chestnut rufous; back and rump more tawny chestnut

♀Black-crested Antshrike, *Thamnophilus radiatus nigricristatus*

Vol. V   p. 34: pileum deep cinnamon-rufous or rufous chestnut

♀Colombian Crested Antshrike, *Hypolophus canadensis pulchellus*

Vol. V   p. 22: tail cinnamon-rufous or rufous chestnut

Tawny Antshrike, *Thamnistes a. anabatina*

Vol. V   p. 20: pileum cinnamon-rufous or rufous chestnut

♀Fasciated Antshrike, *Cymbilaemus lineatus fasciatus*

Vol. IV   p. 765: above dull cinnamon-rufous or hazel

Rufous Manakin, *Laniocera rufescens*

*Vol. IV   pp. 629, 625, 621, 617, 614: inner webs of rectrices cinnamon-rufous (*M. crinitus* specimens 785958 and 369260 averaged 7.0YR 3.7/4.1 by instrument measurement, visual measures were closer to 7.0YR 4.0/6.0)

Nutting's Flycatcher, *Myiarchus n. nuttingi*

Ash-throated Flycatcher, *Myiarchus c. cinerascens*

Mexican Crested Flycatcher, *Myiarchus m. mexicanus*

Ober's Flycatcher, *Myiarchus o. oberi*

Crested Flycatcher, *Myiarchus crinitus*

Vol. IV   p. 187: above dull cinnamon-rufous or tawny rufous

Brown Thrasher, *Toxostoma rufum*

Vol. IV   p. 175: throat, chest deep cinnamon-rufous or chestnut rufous

Martinique Solitaire, *Myadestes g. genibarbis*

Vol. IV   p. 142: below dull cinnamon-rufous or cinnamon chestnut (in spring)

♂ Bluebird, *Sialia s. sialis*

*Vol. IV   p. 97: below deep cinnamon-rufous (in spring) (♀ duller, paler) (specimens 375908 and 375983 averaged 6.0YR 3.3/5.3)

♂ American Robin, *Planesticus m. migratorius*

Vol. III   p. 504: back cinnamon-rufous or dull chestnut, rump dull cinnamon-rufous or cinnamon-brown (in summer)

Hooded Cactus Wren, *Heleodytes c. capistratus*

Vol. III   p. 230: broad eyebrow cinnamon-rufous or rufous chestnut

Costa Rican Pepper-Shrike, *Cyclarhis flavipectus subflavescens*

Vol. II   p. 537: cheek cinnamon-rufous or rufous chestnut (in spring)

♂ Cape May Warbler, *Dendroica tigrina*

Vol. II   p. 527: pileum rich rufous chestnut; rest of head paler, intermediate between orange-rufous and cinnamon-rufous

♂ Panama Yellow Warbler, *Dendroica erithachorides*

Vol. II   p. 526: pileum rufous chestnut; rest of head, throat, upper chest similar but paler, between orange-rufous and cinnamon-rufous

♂ Martinique Yellow Warbler, *Dendroica rufigula*

Vol. II   p. 380: below cinnamon-rufous or russet

♂ Mexican Diglossa, *Diglossa baritula*

*Vol. I   p. 423: sides cinnamon-rufous (specimens 273050 and 761844 averaged 7.0YR 3.5/5.0)

Rufous-sided Towhee, *Pipilio e. erythrophthalmus*

Vol. I   p. 311: pileum cinnamon-rufous to rufous chestnut

Chipping Sparrow, *Spizella s. socialis*

Ferruginous ~ Color 41

Ferruginous is shown in both of Ridgway's color guides. In *Color Standards* it occurs between swatches for Cinnamon-Rufous and Vinaceous-Rufous. All three swatches are nearly identical in my copy, and some deterioration of color is indicated. In the NATURALIST'S COLOR GUIDE Ferruginous has a quality a little ruddier than Cinnamon-Rufous and less intense than Vinaceous-Rufous. It is materially darker and less intense than Ridgway's and Palmer's Rufous.

The COLOR GUIDE swatch chosen for Ferruginous has a 10.0R 5.0/8.0 notation.

Notations of Colors Related to Ferruginous

Color Name	Ridgway Notation	Hamly Notation	Villalobos Notation
Vinaceous-Rufous	XIV 7'i	10.0 R 4.5/ 9.0	SO- 9-7°
Ferruginous	XIV 9'i	1.0 YR 4.5/ 8.0	SO-10-7°
Ferruginous, measure of 1912		2.1 YR 4.8/ 7.2*	
Ferruginous, measure of 1886		1.8 YR 5.0/ 9.2*	
Ferruginous, Guide swatch		10.0 R 5.0/ 8.0*	
Cinnamon-Rufous		(see previous table)	
Rufous	XIV 9'	10.0 R 5.2/10.0	SO-13-9°
Rufous, measure of 1912		0.5 YR 5.7/ 9.0*	
Rufous, measure of 1886		1.7 YR 6.2/ 9.0*	
Rufous, Palmer glossy		10.0 R 5.7/10.4*	
Rufous, Palmer matte		10.0 R 6.0/ 9.1*	

Citations for Ferruginous

Vol. XI p. 226: scapulars, back, rump bright orange-cinnamon to ferruginous (plus dark markings) (light phase)
Ferruginous Rough-legged Hawk, *Buteo regalis*

Olive-Gray ~ Color 42
Grayish Olive ~ Color 43

More than a dozen grayish olive colors are depicted in *Color Standards*. They are primarily distinguished by their grayishness, but all have some olivaceous hue. The olive tones are hardly discernible until they are compared with neutral gray color swatches.

Two such colors with reasonably definite names are Olive-Gray (hyphenated) and Grayish Olive. Others are tints or shades of these two. For example, Ridgway shows Olive-Gray as deep, dark, light, and pale and Grayish Olive as deep, dark, and light. Such modifications should be used in a descriptive sense rather than as specific colors.

In Bulletin 50 Ridgway uses olive-grayish as a descriptive term, apparently distinguished from the color Grayish Olive. In the citations olive-grayish is equated with Olive-Gray.

The COLOR GUIDE swatch chosen for Olive-Gray has a 7.5Y 6.0/2.0 notation. Grayish Olive has a 5.0Y 5.0/2.5 notation.

Notations of Colors Related to Olive-Gray

Color Name	Ridgway Notation	Hamly Notation		Villalobos Notation
Olive-Gray	LI 23''''''b	7.5	Y 5.8/2.0	O-12-1°
Olive-Gray, measure of 1912		0.5	GY 5.5/1.0*?	
Olive-Gray, measure of 1886		4.7	Y 6.6/0.7*?	
Olive-Gray, Guide swatch		7.5	Y 6.0/2.0*	
Grayish Olive	XLVI 21''''	5.0	Y 4.8/2.5	O-10-2°
Grayish Olive, measure of 1912		4.5	Y 5.0/1.8*	
Grayish Olive, Guide swatch		5.0	Y 5.0/2.5*	
Light Grayish Olive	XLVI 21''''b	5.0	Y 5.8/2.0	O-12-2°

Citations for Olive-Gray

Vol. III	p. 570: above olive-gray or grayish olive (more olive in autumn) Gray Flycatcher, *Empidonax griseus*
Vol. III	p. 431: above deep olive-gray or smoke gray (spring and summer) Plumbeous Bush-Tit, *Psaltriparus plumbeus*

116

Vol. III	p. 429: above deep olive-gray Lloyd's Bush-Tit, *Psaltriparus melanotis lloydi*
Vol. III	p. 408: above deep olive-gray or mouse gray Mountain Chickadee, *Penthestes gambeli*
Vol. III	p. 407: back deep olive-gray or mouse gray (rump more olive) Mexican Chickadee, *Penthestes sclateri*
Vol. III	p. 404: back between olive-gray and mouse gray Carolina Chickadee, *Penthestes c. carolinensis*
Vol. III	p. 397: back olive-gray (deeper than color in *Nomenclature*) Chickadee, *Penthestes a. atricapillus*
Vol. III	p. 392: above deep olive-gray; below (not belly) pale olive-gray Wollweber's Titmouse, *Baeolophus w. wollweberi*
Vol. III	p. 197: pileum, back dull olive-gray Stephen's Vireo, *Vireo huttoni stephensi*
Vol. III	p. 194: above deep olive-gray or grayish olive Pale Vireo, *Vireo pallens*
Vol. III	p. 188: above deep olive-gray Gundlach's Vireo, *Vireo gundlachii*
Vol. II	p. 753: above deep olive-gray (= olivaceous mouse gray) Lichtenstein's Warbler, *Basileuterus c. culicivorus*
Vol. II	p. 381: above deep olive-gray or grayish olive; below deep olive-grayish ♀Costa Rican Diglossa, *Diglossa plumbea*
*Vol. II	p. 315: back, rump olive-gray (some streaked) ♀Bullock's Oriole, *Icterus bullockii*
*Vol. II	p. 309: above olive-gray (rump more yellowish olive) ♀Scott's Oriole, *Icterus parisorum*
Vol. II	p. 85: back, rump dull olive-gray (brighter in autumn) ♀Hepatic Tanager, *Piranga hepatica*
Vol. II	p. 71: back light olive or olive-grayish ♀Black-backed Spindalis, *Spindalis z. zena*
Vol. I	p. 251: back smoke gray or olive-gray (plus streaks) Rock Sparrow, *Aimophila ruficeps eremoeca*
Vol. I	p. 214: above olive-grayish (back more olive tinged) (plus streaks) Seaside Sparrow, *Ammodramus m. maritimus*
Vol. I	p. 143: pileum deep gray or olive-gray ♂House Sparrow, *Passer domesticus*
Vol. I	p. 128: above olive-grayish or olive, more olivaceous in winter (plus streaks) ♀Purple Finch, *Carpodacus p. purpureus*
Vol. I	p. 126: above olive-grayish (plus streaks) ♀Cassin's Purple Finch, *Carpodacus cassinii*

Citations for Grayish Olive

Vol. V	p. 93: above grayish olive to light olive Tyrannine Antbird, *Cercomacra t. tyrannina*
Vol. V	p. 85: back, rump deep grayish olive or olive-slaty Northern Long-billed Antwren, *Ramphocaenus r. rufiventris*

117

Vol. V	p. 58: back, rump light grayish olive or hair brown
	Spotted-crowned Antvireo, *Dysithamnus puncticeps*
*Vol. IV	p. 712: back, rump grayish olive
	Thick-billed Kingbird, *Tyrannus crassirostris*
*Vol. IV	p. 621: above grayish olive; nape grayer; tail coverts browner
	Mexican Crested Flycatcher, *Myiarchus m. mexicanus*
*Vol. IV	p. 594: above grayish olive, paler posteriorly (more olive in autumn)
	Phoebe, *Sayornis phoebe*
Vol. IV	p. 570: above grayish olive or olive-gray (more olive in autumn)
	Gray Flycatcher, *Empidonax griseus*
*Vol. IV	p. 521: above olive but grayer (less olive) than *M. p. pertinax;* below more extensively olive-grayish and deeper color than *M. virens*
	Western Wood Pewee, *Myiochanes r. richardsonii*
Vol. IV	p. 518: chest, sides of breast pale grayish olive
	Wood Pewee, *Myiochanes virens*
*Vol. IV	p. 513: above grayish olive (more olive in autumn)
	Swainson's Flycatcher, *Myiochanes p. pertinax*
*Vol. IV	p. 495: pileum, nape, upper back grayish olive
	Fulvous-throated Flycatcher, *Terenotriccus erythrurus fulvigularis*
*Vol. IV	p. 414: above grayish olive (back more olive than pileum) (more olive in autumn)
	Beardless Flycatcher, *Camptostoma imberbe*
Vol. IV	p. 172: above grayish olive or hair brown
	Cuban Solitaire, *Myadestes elizabeth*
Vol. IV	p. 59: above grayish olive or olive; side of head grayish olive (brighter in autumn)
	Gray-cheeked Thrush, *Hylocichla a. aliciae*
Vol. III	p. 194: above grayish olive or deep olive-gray
	Pale Vireo, *Vireo pallens*
*Vol. II	p. 639: above grayish olive (pileum darker)
	Louisiana Water Thrush, *Seiurus motacilla*
*Vol. II	p. 409: upper parts grayish olive
	Mexican Bananaquit, *Coereba mexicana*
Vol. II	p. 381: above grayish olive or deep olive-gray
	♀Costa Rican Diglossa, *Diglossa plumbea*
*Vol. II	p. 66: back grayish olive
	♀Porto Rican Spindalis, *Spindalis portoricensis*
Vol. I	p. 635: head, body red of ♂ replaced by grayish olive or buffy grayish
	♀Cardinal, *Cardinalis c. cardinalis*
Vol. I	p. 478: ♂ back dull grayish olive (rump much lighter); ♀ above light grayish olive
	Parrot Ground-Finch, *Camarhynchus psittaculus*
Vol. I	p. 223: back general color light olive or grayish olive
	Acadian Sharp-tailed Sparrow, *Ammodramus caudacutus subvirgatus*

118

Smoke Gray ~ Colors 44 & 45

Smoke Gray is a color closely related to Grayish Olive, but it has a paler and grayer quality that tends to mask its olive tone. Its quality can be clarified by a comparison of two Ridgway descriptions of closely related flycatchers. Swainson's Flycatcher, *Contopus p. pertinax,* is described as "grayish olive above, in spring plumage; more decidedly olive in autumn." Coues's Flycatcher, *Contopus pertinax pallidiventris,* is described as "deep smoke gray above; chest area lighter smoke gray." An examination of specimens of the two species confirms the color difference between Grayish Olive and Smoke Gray even more clearly than is accomplished by the color swatches.

Palmer (1962) correlates his Smoke Gray swatch with a Villalobos notation of O-13-2°, which is the same as that given by Villalobos for Ridgway's swatch. However, spectrophotometric measurement indicates a slightly browner tone for Palmer's Smoke Gray, with a Munsell notation of 1.5Y 6.0/2.4. It has a somewhat smokier quality and may be preferred by some to Ridgway's color. Despite the risk of creating some confusion, the NATURALIST'S COLOR GUIDE shows both colors under the same name with different numbers.

The COLOR GUIDE swatch chosen for Smoke Gray has a 5.0Y 7.0/2.0 notation. The swatch for Palmer's color has a 2.5Y 6.0/2.4 notation.

Notations of Colors Related to Smoke Gray

Color Name	Ridgway Notation	Hamly Notation	Villalobos Notation
Smoke Gray	XLVI 21''''d	5.0 Y 6.8/2.0	O-13-2°
Smoke Gray, measure of 1912		2.0 Y 6.1/1.2*?	
Smoke Gray, measure of 1886		1.4 Y 6.4/1.2*?	
Smoke Gray, Guide swatch		5.0 Y 7.0/2.0*	
Smoke Gray, Palmer		1.5 Y 6.0/2.4*	O-13-2°
Smoke Gray, Guide swatch		2.5 Y 6.0/2.4*	

Citations for Smoke Gray

Vol. XI   p. 431: head pale smoke gray to smoke gray
           Harpy Eagle, *Harpia harpyja*
Vol. VII   p. 64: above soft smoke gray (pileum browner, between mouse
           gray and hair brown)
           Haitian Lizard-Cuckoo, *Saurothera dominicensis*

Vol. IV	p. 873: above general color between smoke gray and drab-gray
	Gray-headed Tityra, *Tityra semifasciata griseiceps*
Vol. IV	p. 707: above (not tail) about No. 6 gray inclining to smoke brown
	Gray Kingbird, *Tyrannus d. dominicensis*
Vol. IV	p. 515: above deep smoke gray; chest lighter smoke gray (as compared with *M. p. pertinax*)
	Coues's Flycatcher, *Myiochanes pertinax pallidiventris*
Vol. IV	p. 505: above slaty olive or dark smoke gray
	Olive-sided Flycatcher, *Nuttallornis borealis*
Vol. IV	p. 234: above mouse gray, or between that and smoke gray
	Guiana Mockingbird, *Mimus g. gilvus*
Vol. IV	p. 225: above deep smoke gray (but equated with brownish gray)
	Mockingbird, *Mimus p. polyglottos*
Vol. IV	p. 199: above between drab-gray and smoke gray in autumn (browner in spring)
	Curve-billed Thrasher, *Toxostoma c. curvirostre*
Vol. IV	p. 163: above varying drab-gray or smoke gray to hair brown; below similar but paler
	Townsend's Solitaire, *Myadestes townsendi*
Vol. III	p. 432: above deep smoke gray
	Bush-Tit, *Psaltriparus m. minimus*
Vol. III	p. 430: above deep olive-gray or smoke gray (spring and summer)
	Plumbeous Bush-Tit, *Psaltriparus plumbeus*
Vol. III	p. 390: above drab-gray or smoke gray
	Gray Titmouse, *Baeolophus inornatus griseus*
Vol. III	p. 281: nape, back smoke gray or drab-gray (winter)
	Clarke's Nutcracker, *Nucifraga columbiana*
Vol. III	p. 153: pileum, nape light mouse gray or smoke gray
	Warbling Vireo, *Vireosylva g. gilva*
Vol. III	p. 117: head soft smoke gray
	Guatemalan Ptilogonys, *Ptilogonys cinereus molybdophanes*
*Vol. III	p. 116: head (♀) and crest (♂) smoke gray
	Mexican Ptilogonys, *Ptilogonys c. cinereus*
Vol. I	p. 625: head and neck between drab-gray and smoke gray
	Pyrrhuloxia, *Pyrrhuloxia s. sinuata*
Vol. I	p. 395: above mouse gray or deep smoke gray (little, if any, brown)
	Slate-colored Sparrow, *Passerella iliaca schistacea*
*Vol. I	p. 303: throat, chest smoke gray
	Guatemala Junco, *Junco alticola*
Vol. I	p. 291: rump, tail coverts dull No. 6 gray or approaching smoke gray (paler in winter)
	♂ Pink-sided Junco, *Junco mearnsi*
Vol. I	p. 251: back smoke gray or olive-gray (some near ash gray) plus streaks
	Rock Sparrow, *Amophila ruficeps eremoeca*
Vol. I	p. 39: above deep smoke gray (head darker, rump paler)
	♀ Evening Grosbeak, *Hesperiphona v. vespertina*

120

Olive-Green ~ Colors 46, 47 & 48

Olive-Green occurs frequently in the plumage of birds. It is one of the most used color names, not only by Ridgway but by most ornithologists. Ridgway uses it over and over again without modification. Of the eighty-three citations listed, fifty-six (marked with asterisks) describe a color as unequivocally olive-green.

One tends to think of Olive-Green as a definite, uncomplicated color, one readily recognized and confidently applied to a plumage. In many instances this is true, but a study of many of the birds cited indicates a misplaced confidence, perhaps even a degree of looseness in its use. Although many plumages are called olive-green in the citations, examination indicates a wide range of color—some approaching grayish tones, others clearly green tones, and others more yellowish tones.

The NATURALIST'S COLOR GUIDE accepts some of the broad range of color for Olive-Green by including hues ranging from 5.0Y to 10.0Y. The COLOR GUIDE swatch chosen as a basic form of Olive-Green (Color 46) has an 8.5Y 4.0/4.0 notation, which is close to that measured by Hamly for Ridgway's swatch in *Color Standards.* Two auxiliary swatches are also shown to give an idea of the range of the colors. Color 47 has a 10.0Y 4.0/3.5 notation, and Color 48 has a 6.0Y 4.0/3.5 notation. The tabulation includes three Ridgway colors that, while visibly different, are not excessively so: Dark Citrine, Olive-Citrine, and Roman Green.

Ridgway's reliance on Olive-Green is not as complete as might be inferred from the long list of citations. In other descriptions in Bulletin 50 he often modifies it—grayish olive-green, dull or dark olive-green, dull, dusky, or dark greenish olive, and often simply greenish olive. Only Dark Greenish Olive is shown in *Color Standards.* All these modifications are gathered under Greenish Olive (Color 49) in the NATURALIST'S COLOR GUIDE.

One other modification is included (Yellowish Olive-Green, Color 50) in an attempt to eliminate the confusion present in Ridgway's descriptions. There are some citations for it, and they are useful because the birds cited actually indicate a greater difference from Olive-Green than that shown in *Color Standards.*

Notations of Colors Related to Olive-Green

Color Name	Ridgway Notation	Hamly Notation	Villalobos Notation
Dark Citrine	IV 21m	5.0 Y 4.2/4.0	OY-6-5°
Olive-Citrine	XVI 21'm	5.0 Y 3.8/3.5	Y-5-5°
Olive-Green	IV 23m	7.5 Y 4.0/3.5	Y-6-7°
Olive-Green, measure of 1912		9.8 Y 3.9/3.2*	
Olive-Green, measure of 1886		8.0 Y 4.4/4.1*	
Olive-Green, Guide swatch		8.5 Y 4.0/4.0*	
Olive-Green, Guide swatch		10.0 Y 4.0/3.5*	
Olive-Green, Guide swatch		6.0 Y 4.0/3.5*	
Yellowish Olive	XXX 23''k	7.5 Y 4.5/4.0	YYO-7-4°
Yellowish Olive, measure of 1912		7.7 Y 4.3/2.8*	
Roman Green	XVI 23'm	7.5 Y 4.0/3.5	Y-4-5°

Citations for Olive-Green

*Vol. VI   p. 349: above olive-green
Cassin's Araçari, *Selenidera spectabilis*

*Vol. VI   p. 308: above olive-green (nape tinged yellow)
Haitian Piculet, *Nesoctites micromegas*

*Vol. IV   p. 792: above olive-green
Brown-capped Tyrannulet, *Microtriccus brunneicapillus*

*Vol. IV   p. 791: above olive-green
Gray-capped Tyrannulet, *Microtriccus semiflavus*

Vol. IV   p. 768: above bright olive-green
Gray-headed Manakin, *Piprites griseiceps*

*Vol. IV   p. 755: above olive-green
♀Costa Rican White-throated Manakin, *Corapipo leucorrhoa altera*

*Vol. IV   p. 739: above olive-green
♀Long-tailed Manakin, *Chiroprion linearis*

*Vol. IV   p. 737: above olive-green
♀Sharp-tailed Manakin, *Chiroprion lanceolata*

*Vol. IV   p. 732: rump olive-green (♀ above olive-green); below light olive-green
♂Gould's Manakin, *Manacus vitellinus*

*Vol. IV   p. 731: above, including head, olive-green (yellower on lower rump)
♀Cande's Manakin, *Manacus candei*

Vol. IV   p. 665: above greenish olive to almost olive-green
Boat-billed Flycatcher, *Megarynchus pitangua mexicanus*

Vol. IV   p. 582: above bright olive-green
Salvin's Flycatcher, *Empidonax salvini*

*Vol. IV   p. 469: above olive-green
Yellow Flycatcher, *Capsiempis flaveola*

*Vol. IV   p. 467: above olive-green
Yellow-green Leptopogon, *Leptopogon flavovirens*

*Vol. II	p. 743: above olive-green (♀ more yellowish olive-green) ♂ Bell's Warbler, *Basileuterus b. belli*
*Vol. II	p. 710: above olive-green Wilson's Warbler, *Wilsonia p. pusilla* (*W. pusilla pileolata,* more yellowish olive-green)
*Vol. II	p. 628: above olive-green Mourning Warbler, *Oporornis philadelphia*
*Vol. II	p. 626: above olive-green (♀ similar) ♂ Connecticut Warbler, *Oporornis agilis*
*Vol. II	p. 622: above olive-green ♂ Kentucky Warbler, *Oporornis formosa*
Vol. II	p. 599: above bright olive-green (scapulars more grayish) ♂ Pine Warbler, *Dendroica v. vigorsii*
Vol. II	p. 477: above bright olive-green (♀ duller) ♂ Hartlaub's Warbler, *Oreothlypis superciliosa*
*Vol. II	p. 468: above olive-green (brightest on rump) (♀ duller) ♂ Nashville Warbler, *Helminthophila r. rubricapilla*
Vol. II	p. 466: above bright olive-green Lutescent Warbler, *Helminthophila celata lutescens*
*Vol. II	p. 460: above olive-green (brightest on rump) ♂ Tennessee Warbler, *Helminthophila peregrina*
*Vol. II	p. 458: above olive-green ♂ Bachman's Warbler, *Helminthophila bachmani*
Vol. II	p. 455: above bright olive-green (rump more yellowish) (♀ duller) ♂ Blue-winged Warbler, *Helminthophila pinus*
Vol. II	p. 452: above bright olive-green (♀ duller) ♂ Lawrence's Warbler, *Helminthophila lawrencii*
Vol. II	p. 387: above olive-green or dull green; below (partly) pale olive-greenish ♀ Blue Honeycreeper, *Cyanerpes cyaneus*
*Vol. II	p. 167: above olive-green Drab-breasted Chlorospingus, *Chlorospingus hypophaeus*
*Vol. II	p. 165: above olive-green Sooty-capped Chlorospingus, *Chlorospingus pileatus*
*Vol. II	p. 163: above olive-green Carmiol's Chlorospingus, *Chlorospingus albitempora*
*Vol. II	p. 161: above clear olive-green Brown-headed Chlorospingus, *Chlorospingus ophthalmicus*
Vol. II	p. 155: above bright olive-green Carmiol's Tanager, *Chlorothraupis carmioli*
Vol. II	p. 154: above bright olive-green Yellow-browed Tanager, *Chlorothraupis olivaceus*
*Vol. II	p. 66: above olive-green (yellower posteriorly) ♂ Porto Rican Spindalis, *Spindalis portoricensis*
*Vol. II	p. 31: above olive-green (yellower posteriorly) ♀ Jamaican Euphonia, *Pyrrhuphonia jamaica*
*Vol. II	p. 29: above olive-green plus metallic sheen (neck, chest yellowish olive-green) ♂ Gould's Euphonia, *Euphonia gouldi*

124

*Vol. II	p. 20: above olive-green, plus bluish green gloss (rump yellower)
	♀Cabanis' Euphonia, *Euphonia gracilis*
Vol. II	p. 16: above lighter olive-green (lacking bluish green gloss)
	♀Green Euphonia, *Euphonia flavifrons*
*Vol. II	pp. 12, 14, 15, 18, 26: above olive-green plus metallic bluish green glosses
	♀Euphonias, *Euphonia elegantissimia, E. musica, E. sclateri, E. fulvicrissa, E. herundinacea*
*Vol. I	p. 669: above (not rump) olive-green
	Panama Streaked Saltator, *Saltator albicollis isthmicus*
*Vol. I	p. 658: above olive-green (nape yellower)
	♀Crimson-collared Grosbeak, *Rhodothraupis celaeno*
*Vol. I	p. 546: above olive-green
	♂Yellow-shouldered Grassquit, *Loxipasser anoxanthus*
Vol. I	p. 538: above more olive-greenish than *Euetheia b. bicolor*
	Carib Grassquit, *Euetheia bicolor omissa*
*Vol. I	p. 536: above olive-green
	Mexican Grassquit, *Euetheia canora*
*Vol. I	p. 472: above olive-green
	Large-footed Sparrow, *Pezopetes capitalis*
*Vol. I	p. 468: above olive-green
	Gray-striped Buarremon, *Buarremon assimilis*
*Vol. I	p. 467: above olive-green
	Green-striped Buarremon, *Buarremon virenticeps*
*Vol. I	p. 465: above olive-green
	Chestnut-capped Buarremon, *Buarremon brunneinuchus*
Vol. I	p. 458: above general color deep olive-green
	Barranca Sparrow, *Lysurus crassirostris*
Vol. I	p. 453: above bright olive-green
	Richmond's Sparrow, *Arremonops conirostris richmondi*
Vol. I	p. 452: above bright olive-green
	Green-backed Sparrow, *Arremonops chloronotus*
Vol. I	p. 449: above olive-greenish
	Nicoya Sparrow, *Arremonops s. superciliosus*
*Vol. I	p. 407: above olive-green
	Collared Towhee, *Pipilo t. torquatus*
*Vol. I	p. 114: above olive-green (♀ olive-greenish)
	♂Arkansas Goldfinch, *Astragalinus p. psaltria*
*Vol. I	p. 104: above olive-green (rump lighter)
	♀Bryant's Siskin, *Spinus xanthogaster*
Vol. I	p. 44: back, scapulars light olive-green
	♂Abeille's Grosbeak, *Hesperiphona abeillii*

Greenish Olive ~ Color 49

Greenish Olive is a color term frequently used by Ridgway in Bulletin 50, but not identified by a color swatch in *Color Standards*. He associates it more often with dull olive-green than with any other color (see the citations). He also uses the term grayish olive-green with equal or greater frequency, but again there is no swatch in *Color Standards*. Other terms used but not shown are dull and dusky greenish olive and dull and dark olive-green.

These terms have all been grouped in the NATURALIST'S COLOR GUIDE under the name Greenish Olive, which is close to Ridgway's Dark Greenish Olive swatch in *Color Standards*.

The COLOR GUIDE swatch chosen for Greenish Olive has a 7.5Y 3.5/3.0 notation.

Notations of Colors Related to Greenish Olive

Color Name	Ridgway Notation	Hamly Notation	Villalobos Notation
Dark Greenish Olive	XXX 23″m	7.5 Y 3.8/3.0	YYO-5-2°
Dark Greenish Olive, measure of 1912		8.9 Y 3.5/1.9*?	
Greenish Olive, Guide swatch		7.5 Y 3.5/3.0*	

Citations for Dark Greenish Olive

*Vol. VI   p. 345: back dark greenish olive (plus gloss)
            Frantzius' Araçari, *Pteroglossus frantzii*

Citations for Dull or Dusky Greenish Olive

*Vol. II   p. 494: pileum, nape dull greenish olive (to yellowish olive-green) (♀ in summer)
            Olive Warbler, *Peucedramus olivaceus*
*Vol. II   p. 397: pileum, nape dull or dusky greenish olive
            ♀Scarlet-thighed Dacnis, *Dacnis venusta*

Citations for Greenish Olive

Vol. IV   p. 883: above dull olive-green or greenish olive
           ♀Costa Rican Bellbird, *Procnias tricarunculata*
Vol. IV   p. 746: above dull olive-green or greenish olive
           ♀Red-capped (yellow-thighed) Manakin, *Pipra m. mentalis*
Vol. IV   p. 558: above more greenish olive (than other empidonaces)
           Alder Flycatcher, *Empidonax traillii alnorum*

126

Vol. IV	p. 549: above dull olive green or greenish olive
	Yellow-bellied Flycatcher, *Empidonax flaviventris*
Vol. IV	p. 501: above bright greenish olive
	Yellow-vented Flycatcher, *Mitrephanes aurantiiventris*
*Vol. IV	p. 490: above greenish olive
	Sulphur-rumped Myiobius, *Myiobius xanthopygus sulphur-eipygius*
*Vol. IV	p. 488: above greenish olive
	Black-tailed Myiobius, *Myiobius barbatus atricaudus*
Vol. IV	p. 461: above dull olive-green or greenish olive
	Olivaceous Mionectes, *Mionectes o. olivaceus*
*Vol. IV	p. 454: above greenish olive
	Mexican Pipromorpha, *Pipromorpha a. assimilis*
Vol. IV	p. 417: above light greenish olive
	Yellow-bellied Camptostoma, *Camptostoma pusillum flaviventre*
Vol. III	p. 204: above dull olive-green or greenish olive (rump brighter)
	Bell's Vireo, *Vireo b. bellii*
Vol. III	p. 199: above dull olive-green or greenish olive
	Carmiol's Vireo, *Vireo carmioli*

Citations for Dull and Dark Olive-Green

See citations for Greenish Olive (Vol. IV, pp. 883, 746, 549, 461; Vol. III, pp. 204, 199).

*Vol. IV	p. 662: above dull olive-green
	Golden-bellied Flycatcher, *Myiodynastes hemichrysus*
Vol. II	p. 215: back dark olive-green or dull bottle-green
	♂ Florida Grackle, *Quiscalus quiscula oglaeus*
*Vol. II	p. 25: above dull olive-green (rump brighter)
	♀ Godman's Euphonia, *Euphonia godmani*
*Vol. I	p. 537: above dull olive-green
	Bahama Grassquit, *Euethiea b. bicolor*
*Vol. I	p. 100: above dark olive-green (paler below)
	♂ Guatemalan Pine-Siskin, *Spinus atriceps*

Citations for Grayish Olive-Green

Vol. IV	p. 794: above grayish olive-green (or olive-green)
	Riker's Yellow-crowned Tyrannulet, *Tyrannus elatus reguloides*
*Vol. IV	p. 669: above grayish olive-green
	White-ringed Flycatcher, *Coryphotriccus albovittatus*
*Vol. IV	p. 533: above grayish olive-green
	Green-crested Flycatcher, *Empidonax virescens*
Vol. IV	p. 374: above grayish olive-green (but varies to buffy olive-green)
	Pygmy Flycatcher, *Atalotriccus p. pilaris*
*Vol. III	p. 230: above grayish olive-green
	Cozumel Pepper-Shrike, *Cyclarhis insularis*
Vol. III	p. 225: back dull grayish olive-green
	♀ DuBus' Shrike-Vireo, *Vireolanius melitophrys*

*Vol. III	p. 151: above grayish olive-green
	Philadelphia Vireo, *Vireosylva philadelphica*
*Vol. III	p. 147: above grayish olive-green
	Red-eyed Vireo, *Vireosylva olivacea*
*Vol. II	p. 745: above grayish olive-green
	Rufous-capped Warbler, *Basileuterus rufifrons*
*Vol. II	p. 692: above grayish olive-green, lower rump grayer (spring and summer)
	Yellow-breasted Chat, *Icteria v. virens*
Vol. II	p. 687: back, scapulars dull grayish olive-green (spring and summer)
	Ralph's Ground-Chat, *Chamaethlypis p. poliocephala*
Vol. II	p. 661: dull grayish olive-green (spring and summer) (less gray in fall)
	♂ Maryland Yellow-throat, *Geothlypis t. trichas* (Northern Yellow-throat, *G. t. brachidactyla,* is more decidedly olive-green)
*Vol. II	p. 570: above grayish olive-green (to light bluish gray)
	♀Cerulean Warbler, *Dendroica rara*
*Vol. II	p. 462: above grayish olive-green (rump more yellowish olive-green)
	♂ Orange-crowned Warbler, *Helmitheros c. celata*
*Vol. II	p. 458: tail coverts grayish olive-green
	♂ Bachman's Warbler, *Helmitheros bachmani*
*Vol. II	p. 439: above grayish olive-green (less gray in winter)
	Worm-eating Warbler, *Helmitheros vermivorus*
Vol. II	p. 271: back dull grayish olive-green (immature)
	Cuban Oriole, *Icterus hypomelas*
*Vol. II	p. 102: above grayish olive-green (grayer posteriorly)
	♂ Red-headed Tanager, *Piranga erythrocephala*
*Vol. II	p. 22: above grayish olive-green (lower rump yellower)
	♀Lesson's Euphonia, *Euphonia affinis*
*Vol. I	p. 447: above grayish olive-green
	Texas Sparrow, *Arremonops r. rufivirgatus*
*Vol. I	p. 45: back, rump grayish olive greenish
	♀Abeille's Grosbeak, *Hesperiphona abeillii*

Yellowish Olive-Green ~ Color 50

Yellowish Olive-Green is not identified by a swatch in *Color Standards,* which is regrettable because Ridgway uses the term frequently in Bulletin 50. To judge from his descriptions, he clearly had a specific color quality in mind; he is not merely modifying Olive-Green by calling it yellowish, although the omission of a swatch encourages that thought. There are some sixty citations, of which about forty are not modified and about fifteen only slightly qualified as light, deep, or bright yellowish olive-green. A few of the citations relate it to Olive-Yellow (Color 52), which is a much brighter color.

In order to arrive at an acceptable notation for Yellowish Olive-Green, it was necessary to examine many of the plumages described in Bulletin 50. As has often occurred during such investigations, a confusing variety of colors was found. Although called yellowish olive-green, many seemed more clearly olive-yellow, olive-green, or even plain olive. Spectrophotometric measurement of the specimens that *were* yellowish olive-green resulted in a Munsell notation for the NATURALIST'S COLOR GUIDE.

The tabulation consists of all the colors in the COLOR GUIDE of an olivaceous nature. They are listed with the notations chosen for the guide. If they also meet Hamly's notations for Ridgway's swatches, they are marked with a double asterisk.

The COLOR GUIDE swatch chosen for Yellowish Olive-Green has an 8.0Y 4.5/6.0 notation.

Notations of Colors Related to Olivaceous

Color Number and Name		Ridgway Notation	Hamly Notation		Villalobos Notation
28	Olive-Brown	XL 17'''k	10.0 YR	4.0/2.0**	O- 6-3°
29	Brownish Olive	XXX 19''m	2.5 Y	4.0/2.0*	O- 5-2°
30	Olive	XXX 21''m	4.5 Y	4.0/2.0*	OOY- 5-2°
42	Olive-Gray	LI 23'''''b	7.5 Y	6.0/2.0**	O-12-1°
43	Grayish Olive	XLVI 21''''	5.0 Y	5.0/2.5**	O-10-2°
44	Smoke Gray	XLVI 21''''d	5.0 Y	7.0/2.0**	O-13-2°
46	Olive-Green	IV 23m	8.5 Y	4.0/4.0*	Y- 6-7°
49	Greenish Olive	XXX 23''m	7.5 Y	3.5/3.0*	YYO- 5-2°
50	Yellowish Olive-Green		8.0 Y	4.5/6.0*	
51	Citrine	IV 21k	6.3 Y	5.0/5.0*	YYO- 9-8°
52	Olive-Yellow	XXX 23''	7.5 Y	7.0/7.0**	Y-13-8°

Citations for Yellowish Olive-Green

*Vol. IV p. 733: below yellowish olive-green, becoming olive-yellow on center of belly
    ♀Gould's Manakin, *Manacus vitellinus*

*Vol. IV p. 705: above yellowish olive-green
    Couch's Kingbird, *Tyrannus melancholicus couchii*

Vol. IV p. 700: above mixed yellowish olive-green and gray; yellower in Yucatan; mostly grayish in Cozumel
    Lichtenstein's Kingbird, *Tyrannus melancholicus satrapa*

Vol. IV p. 583: above deep yellowish olive-green
    Yellowish Flycatcher, *Empidonax flavescens*

*Vol. IV p. 394: above yellowish olive-green
    Gray-headed Flycatcher, *Rhynchocyclus cinereiceps*

Vol. IV p. 391: above bright yellowish olive-green
    Yellow-olive Flycatcher, *Rhynchocyclus flavo-olivaceus*

Vol. IV p. 387: above light yellowish olive-green
    Equinoctial Flycatcher, *Craspedoprion aequinoctialis*

*Vol. IV p. 366: above yellowish olive-green
    Black-headed Tody-Flycatcher, *Todirostrum nigriceps*

*Vol. III p. 221: above yellowish olive-green
    Yellow-green Pachysylvia, *Pachysylvia viridiflava*

*Vol. III p. 220: above yellowish olive-green
    Lawrence's Pachysylvia, *Pachysylvia a. aurantiifrons*

*Vol. III p. 216: above yellowish olive-green
    Gray-headed Greenlet, *Pachysylvia decurtata*

*Vol. II p. 705: above yellowish olive-green
    Hooded Warbler, *Wilsonia mitrata*

*Vol. II p. 607: above yellowish olive-green
    ♂Prairie Warbler, *Dendroica discolor*

Vol. II p. 515: above bright yellowish olive-green
    ♂Jamaican Yellow Warbler, *Dendroica p. petechia*

*Vol. II p. 508: above yellowish olive-green
    ♂Yellow Warbler, *Dendroica a. aestiva*

*Vol. II p. 494: ♂ nape and extreme upper back yellowish olive-green (in summer), ♀ pileum and nape varying yellowish olive-green to dull greenish olive
    Olive Warbler, *Peucedramus olivaceus*

Vol. II p. 471: rump yellowish olive-green or olive-yellow
    ♂Virginia's Warbler, *Helminthophila virginiae*

*Vol. II p. 442: back, scapulars yellowish olive-green
    Prothonotary Warbler, *Protonotaria citrea*

Vol. II p. 409: rump light yellowish olive-green or olive-yellow
    Mexican Bananaquit, *Coereba mexicana*

*Vol. II p. 303: above yellowish olive-green
    Jamaican Oriole, *Icterus leucopteryx*

*Vol. II p. 275: above yellowish olive-green
    ♀Orchard Oriole, *Icterus spurius*

*Vol. II p. 166: above yellowish olive-green
    Yellow-breasted Chlorospingus, *Chlorospingus punctulatus*

130

*Vol. II   p. 140: above yellowish olive-green
            Gray-headed Tanager, *Eucometis s. spodocephala*
*Vol. II   p. 139: above yellowish olive-green
            Gray-crested Tanager, *Eucometis cristata*
*Vol. II   p. 136: above yellowish olive-green
            ♀Veraguan White-shouldered Tanager, *Tachyphonus nitidis-
            simus*
*Vol. II   p. 134: above yellowish olive-green
            ♀Costa Rican White-shouldered Tanager, *Tachyphonus axil-
            laris*
Vol. II   p. 133: above bright yellowish olive-green
            ♀White-shouldered Tanager, *Tachyphonus lactuosus*
*Vol. II   p. 128: above yellowish olive-green
            ♂Gray-crowned Palm Tanager, *Phaenicophilus poliocephalus*
*Vol. II   p. 127: above yellowish olive-green
            Palm Tanager, *Phaenicophilus palmarum*
*Vol. II   p. 106: above yellowish olive-green
            ♀Black and Yellow Tanager, *Hemithraupis chrysomelas*
Vol. II   p. 102: above bright yellowish olive-green
            ♂Red-headed Tanager, *Piranga erythrocephala*
*Vol. II   p. 87: above yellowish olive-green
            ♀Brick-red Tanager, *Piranga t. testacea*
Vol. II   p. 79: above yellowish olive-green, some with grayish tinge
            ♀Summer Tanager, *Piranga r. rubra*
Vol. II   p. 73: back deep yellowish olive-green
            ♂Cozumel Spindalis, *Spindalis benedicti*
*Vol. II   p. 69: back yellowish olive-green
            ♂Cuban Spindalis, *Spindalis pretrei*
*Vol. II   p. 67: back yellowish olive-green
            ♂Haitian Spindalis, *Spindalis multicolor*
*Vol. II   p. 64: above yellowish olive-green (yellower posteriorly)
            Jamaican Spindalis, *Spindalis nigricephala*
*Vol. II   p. 29: neck, chest yellowish olive-green
            ♂Gould's Euphonia, *Euphonia gouldi*
*Vol. II   p. 28: above yellowish olive-green
            ♀Thick-billed Euphonia, *Euphonia crassirostris*
Vol. II   p. 24: above deep yellowish olive-green
            ♀White-vented Euphonia, *Euphonia minuta humilis*
*Vol. II   p. 21: above yellowish olive-green
            ♀Yellow-crowned Euphonia, *Euphonia luteicapilla*
Vol. II   p. 20: below yellowish olive-green (or deep olive-yellowish)
            ♀Cabanis' Euphonia, *Euphonia gracilis*
Vol. II   p. 18: breast bright yellowish olive-green
            ♀Fulvous-vented Euphonia, *Euphonia fulvicrissa*
*Vol. II   p. 17: above yellowish olive-green
            ♀Tawny-capped Euphonia, *Euphonia annae*
Vol. II   p. 16: rump yellowish olive-green (or bright olive-yellowish);
            below yellowish olive green
            ♂Green Euphonia, *Euphonia flavifrons*
*Vol. II   p. 15: rump yellowish olive-green (below deep olive-yellow)
            ♀Porto Rican Euphonia, *Euphonia sclateri*

131

Vol. II	p. 14: below light yellowish olive-green or deep olive yellow
	♀ Haitian Euphonia, *Euphonia musica*
Vol. II	p. 12: below light yellowish olive-green (medially more olive yellowish)
	♀ Blue-hooded Euphonia, *Euphonia elegantissima*
Vol. I	p. 663: above bright yellowish olive-green
	Buff-throated Saltator, *Saltator m. magnoides*
Vol. I	p. 661: above bright yellowish olive-green (tail brightest, rich olive-yellow)
	Black-headed Saltator, *Saltator a. atriceps*
*Vol. I	p. 656: back, scapulars yellowish olive-green
	Lesser Bishop Grosbeak, *Caryothraustes poliogaster scapularis*
*Vol. I	p. 622: above yellowish olive-green (more grayish on back, yellower on head and rump)
	♀ Vigor's Grosbeak, *Pheuticus chrysopeplus*
*Vol. I	p. 523: nape, lower back yellowish olive-green (plus streaks) ·
	♂ Mexican Yellow Finch, *Sicalis chrysops*
*Vol. I	p. 107: above yellowish olive-green (rump olive-yellow)
	♂ Haitian Goldfinch, *Loxinutris dominicensis*
*Vol. I	p. 102: above yellowish olive-green (♀ duller)
	♂ Black-headed Siskin, *Spinus n. notatus*

Citrine ~ Color 51

Citrine and a very similar Warbler Green are shown in *Color Standards* but not in the earlier *Nomenclature*. They are of a moderately dark shade and a yellowish green hue. The greenish quality may derive as much from grayishness as from a truly green pigmentation. Ridgway's Warbler Green, for example, was produced from 100 percent yellow pigment, in a blend of 29.5 percent color and 70.5 percent black. No green was used, although greenishness is apparent to the eye. The Munsell notations can also be confusing in this respect, as is indicated by the hue notation of 5.0Y for Citrine—squarely in the center of the yellowish hues rather than green ones. The 7.5Y notation for Warbler Green indicates a slightly greater greenish trend. The Villalobos notations give no hint of greenishness either—the OY, YYO, and Y notations are all in yellow or orange-yellow areas. However, the darker areas of the Villalobos charts are visually greenish. This anomalous condition is especially characteristic of darkened yellowish colors and results in colors seeming olivaceous.

The approximate average of Citrine and Warbler Green was found for the notation for the NATURALIST'S COLOR GUIDE. Some people may prefer to use the name Warbler Green, which has perhaps a more ornithological flavor.

The COLOR GUIDE swatch chosen for Citrine has a 6.3Y 5.0/5.0 notation.

Notations of Colors Related to Citrine

Color Name	Ridgway Notation	Hamly Notation	Villalobos Notation
Citrine	IV 21k	5.0 Y 4.5/5.0	YYO- 9- 8°
Citrine, measure of 1912		5.0 Y 4.7/4.9*	
Citrine, Guide swatch		6.3 Y 5.0/5.0*	
Warbler Green	IV 23k	7.5 Y 5.0/5.0	Y- 9-10°
Dull Citrine	XVI 21'k	5.0 Y 4.5/4.5	Y- 8- 6°
Citrine-Drab	XL 21'''i	5.0 Y 5.2/4.0	OY- 8- 4°
Buffy Olive	XXX 21''k	5.0 Y 5.0/4.0	OOY-11- 4°

Citations for Citrine

No useful citations for Citrine were found in Bulletin 50.

Olive-Yellow ~ Color 52

Olive-Yellow is shown in *Color Standards* and is often used in Bulletin 50. Ridgway frequently uses the descriptive term olive-yellowish—not identified by a swatch in *Color Standards*. The two names seem to be used interchangeably when a description is contrasted with some third color. I have chosen to consider them synonyms and the citations are listed together.

The descriptions often include Light Yellowish Olive, which is a slightly duller color shown in *Color Standards*. The citations seem to distinguish it from Olive-Yellow consistently, although the color difference is not substantial. Olive-Yellow is also distinguished from Yellowish Olive-Green, which is not shown in *Color Standards* but is Color 50 in the NATURALIST'S COLOR GUIDE.

Yellowish Citrine, shown in *Color Standards,* is measured by Hamly equal to Light Yellowish Olive but by Villalobos close to Olive-Yellow. No useful citations were found for it.

Because the amount of colors and citations is limited, the name Olive-Yellow is sufficient and can be modified when necessary.

The COLOR GUIDE swatch chosen for Olive-Yellow has a 7.5Y 7.0/7.0 notation—practically the equivalent of that given by Hamly to the Ridgway swatch.

Notations of Colors Related to Olive-Yellow

Color Name	Ridgway Notation	Hamly Notation	Villalobos Notation
Olive-Yellow	XXX 23″	7.5 Y 6.8/7.0	Y-13- 8°
Olive-Yellow, measure of 1912		8.5 Y 6.4/6.4*	
Olive-Yellow, measure of 1886		8.7 Y 8.0/7.0*	
Olive-Yellow, Guide swatch		7.5 Y 7.0/7.0*	
Light Yellowish Olive	XXX 23″i	7.5 Y 5.8/6.0	YYO-10- 5°
Yellowish Citrine	XVI 23′i	7.5 Y 6.0/6.0	Y-11-10°

Citations for Olive-Yellow

The olive-yellowish of the citations is not a Ridgway color but is merely a descriptive term.

*Vol. VII   p. 171: belly olive-yellowish (clear olive-yellow to dull wax yellow)
         Aztec Paroquet, *Eupsittula astec*

Vol. VII	p. 172: under tail light yellowish olive, changing to deep olive-yellowish
	Aztec Paroquet, *Eupsittula astec*
Vol. VII	p. 169: under tail light yellowish olive, changing to deep olive-yellowish
	Petz's Paroquet, *Eupsittula canicularis*
Vol. VII	p. 168: under tail light yellowish olive, changing to deep olive-yellowish
	Veragua Paroquet, *Eupsittula ocularis*
Vol. VII	p. 164: under tail light yellowish olive, changing to bright olive-yellowish
	Curaçao Paroquet, *Eupsittula p. pertinax*
Vol. VII	p. 160: under wings and tail light yellowish olive, changing to deep olive-yellowish
	Cuban Paroquet, *Aratinga euops*
Vol. VII	p. 159: under wings and tail light yellowish olive, changing to deep olive-yellowish
	Socorro Paroquet, *Aratinga brevipes*
Vol. VII	p. 157: under wings and tail light yellowish olive, changing to deep olive-yellowish
	Green Paroquet, *Aratinga h. holochlora*
Vol. VII	p. 156: under wings and tail light yellowish olive or deep olive-yellowish
	Red-throated Paroquet, *Aratinga rubritorquis*
*Vol. IV	p. 733: below yellowish olive-green, becoming olive-yellow central belly
	♀ Gould's Manakin, *Manacus vitellinus*
Vol. IV	p. 697: sides and flanks olive-yellow or light yellowish olive
	Arkansas Kingbird, *Tyrannus verticalis*
Vol. II	p. 471: rump olive-yellow or yellowish olive-green (♀ duller)
	♂ Virginia's Warbler, *Helminthophila virginiae*
Vol. II	p. 466: below bright olive-yellow
	Lutescent Warbler, *Helminthophila celata lutescens*
Vol. II	p. 421: lower rump olive-yellowish or light yellowish olive
	Curaçao Bananaquit, *Coereba uropygialis*
Vol. II	p. 420: lower rump olive-yellowish or light yellowish olive
	Barbados Bananaquit, *Coereba barbadensis*
Vol. II	p. 415: rump olive-yellowish or light yellowish olive
	St. Vincent Bananaquit, *Coereba saccharina*
Vol. II	p. 409: rump olive-yellow or light yellowish olive-green
	Mexican Bananaquit, *Coereba mexicana*
*Vol. II	p. 408: rump olive-yellow
	San Miguel Bananaquit, *Coereba cerinoclunis*
Vol. II	p. 30: below deep olive-yellowish
	♀ Gould's Euphonia, *Euphonia gouldi*
Vol. II	p. 25: breast pale olive-yellowish
	♀ Godman's Euphonia, *Euphonia godmani*
Vol. II	p. 24: breast deep olive-yellow
	♀ White-vented Euphonia, *Euphonia minuta humilis*
Vol. II	p. 22: below olive-yellowish (or light grayish yellow)
	♀ Lesson's Euphonia, *Euphonia affinis*

Vol. II	p. 20: below deep olive-yellowish (or yellowish olive-green)
	♀ Cabanis' Euphonia, *Euphonia gracilis*
Vol. II	p. 16: rump bright olive-yellow (or yellowish olive-green)
	♂ Green Euphonia, *Euphonia flavifrons*
Vol. II	p. 15: below deep olive-yellow
	♀ Porto Rican Euphonia, *Euphonia sclateri*
Vol. II	p. 14: below deep olive-yellow (or light yellowish olive-green)
	♀ Haitian Euphonia, *Euphonia musica*
Vol. I	p. 661: tail rich olive-yellow (some lights)
	Black-headed Saltator, *Saltator a. atriceps*
*Vol. I	p. 587: breast olive-yellowish (belly yellow)
	♀ Painted Bunting, *Cyanospiza ciris*
*Vol. I	p. 523: sides head and neck olive-yellow
	♂ Mexican Yellow Finch, *Sicalis chrysops*
*Vol. I	p. 107: rump olive-yellow
	♂ Haitian Goldfinch, *Loximitris dominicensis*
*Vol. I	p. 44: breast olive-yellow
	♂ Abeille's Grosbeak, *Hesperiphona abeillii*

Citations for Light Yellowish Olive

See citations for Olive-Yellow (Vol. VII, pp. 172, 169, 168, 164, 160, 159, 157, 156; Vol. IV, p. 697; Vol. II, pp. 421, 420, 415).

*Vol. IV	p. 700: chest light yellowish olive
	Lichtenstein's Kingbird, *Tyrannus melancholicus satrapa*
Vol. IV	p. 697: above light yellowish olive or yellowish gray
	Arkansas Kingbird, *Tyrannus verticalis*
*Vol. II	p. 409: flanks light yellowish olive
	Mexican Bananaquit, *Coereba mexicana*
*Vol. I	p. 45: breast light yellowish olive
	♀ Abeille's Grosbeak, *Hesperiphona abeillii*

136

Buff-Yellow ~ Color 53

Buff-Yellow, a hyphenated name, is one of the twenty-eight colors in *Color Standards* with buff in the color name. All but two (Buff, Color 24, and Buff-Yellow) have been omitted from the NATURALIST'S COLOR GUIDE, partly because the descriptions using them in Bulletin 50 refer to such small areas as to make identification difficult and partly because the color names are largely self-descriptive, merely calling for some buffiness of other colors.

The descriptions refer to such areas as the chin, wingbars, coverts, and feather edges. For example, the kingfisher *Chloroceryle inda* is described in Vol. VI, p. 434, as having the forehead minutely flecked with pale brownish buffy, the chin pale orange-buff or white tinged with orange-buff, the throat feathers tipped with deeper orange-buff. Another kingfisher, *Chloroceryle a. aenea,* is described in Vol. VI, p. 437, as having the outer webs of the secondaries with spots of buff or buffy white, a supraloral spot of ochraceous-buff, the chin and throat fading to orange-buff, the under wing-coverts ochraceous-buff or orange-buff, and the inner webs of the secondaries pale buff. Of course, these descriptions are useful guides in a taxonomic study; they add to the list of facts distinguishing one species from another. But they are not much help in identifying a specific color.

Palmer (1962) shows a swatch called Buffy Yellow; its measurements match very closely with those of Ridgway's Buff-Yellow.

The COLOR GUIDE swatch chosen for Buff-Yellow has a 2.5Y 8.0/6.0 notation.

Notations of Colors Related to Buff-Yellow

Color Name	Ridgway Notation	Hamly Notation	Villalobos Notation
Buff-Yellow	IV 19d	2.5 Y 8.0/7.0	OOY-18-10°
Buff-Yellow, measure of 1912		2.5 Y 8.1/7.0*	
Buffy Yellow, Palmer glossy		3.0 Y 7.9/6.4*	OOY-18-10°
Buffy Yellow, Palmer matte		2.6 Y 8.0/5.7*	
Buff-Yellow, Guide swatch		2.5 Y 8.0/6.0*	
Maize Yellow	IV 19f	2.5 Y 8.5/6.0	OOY-19-12°
Apricot Yellow	XVI 19b	2.5 Y 7.8/9.5	OOY-18-12°
Naples Yellow	XVI 19'd	3.5 Y 8.2/7.0	OY-17- 8°
Naples Yellow, Winsor & Newton		3.0 Y 8.5/6.7*	
Light Orange-Yellow	III 17d	1.0 Y 7.5/9.0	OOY-17-10°
Pale Orange-Yellow	III 17f	1.0 Y 8.0/8.0	OOY-18-10°

Citations for Buff-Yellow

Vol. VI   p. 463: below pale buffy greenish, tinged on chest with yellow-ish buffy
Mexican Motmot, *Momotus m. mexicanus*

Vol. VI   p. 305: belly light buffy yellow (nearly straw or naples yellow)
Panama Piculet, *Picumnus olivaceus panamensis*

Vol. IV   p. 589: throat, belly light yellowish buff or buff-yellow (in autumn); chest, etc., tawny buff
Buff-breasted Flycatcher, *Empidonax fulvifrons pygmaeus*

*Vol. IV   p. 583: edge of wing buff-yellow
Yellowish Flycatcher, *Empidonax flavescens*

Vol. IV   p. 581: edge of wing buffy or ochraceous yellow
Baird's Flycatcher, *Empidonax difficilis bairdi*

Vol. IV   p. 543: below anteriorly pale yellowish buff, to pale buffy yellow posteriorly
♂ Bran-colored Flycatcher, *Myiophobus fasciatus furfurosus*

Vol. IV   p. 501: belly, flanks pale buffy yellow (nearly naples yellow)
Yellow-vented Flycatcher, *Mitrephanes aurantiiventris*

Vol. IV   p. 459: breast, belly light buff-yellow (plus streaks)
♀ Cherrie's Pipromorpha, *Pipromorpha semischistacea*

Vol. IV   p. 354: lower rump, tail coverts buff-yellow or yellowish buff
Mexican Royal Flycatcher, *Onychorhynchus m. mexicanus*

Citations for Naples Yellow

*Vol. IV   p. 804: lower rump, tail coverts naples yellow
♀ Gray-headed Attila, *Attila tephrocephalus*

Vol. IV   p. 379: below nearly naples yellow
♂ Lawrence's Spadebilled Flycatcher, *Placostomus superciliaris*

*Vol. III   p. 221: below naples yellow
Yellow-green Pachysylvia, *Pachysylvia viridiflava*

*Vol. II   p. 98: throat naples yellow
♀ Rose-throated Tanager, *Piranga r. roseogularis*

138

Cream Color ~ Color 54

Cream Color is shown in both *Color Standards* and *Nomenclature*. Palmer (1962) also shows a Cream Color swatch; it seems to have a very slight greenish cast. Ridgway shows other colors close to Cream Color, but they are mostly of a more buffy nature.

It seems obvious that Cream Color must be based on the color of cream, which in days past floated to the top of a bottle of milk. It is impossible to find in today's homogenized or skim milk. Even a bottle of today's cream, though somewhat richer in color than milk, seems to lack the true color of cream.

The COLOR GUIDE swatch chosen for Cream Color has a 3.5Y 8.5/ 4.0 notation.

Notations of Colors Related to Cream Color

Color Name	Ridgway Notation	Hamly Notation	Villalobos Notation
Light Buff	XV 17'f	1.0 Y 8.5/4.5	O-19-6°
Warm Buff	XV 17'd	1.0 Y 7.8/6.0	O-17-7°
Cartridge Buff	XXX 19''f	2.5 Y 8.5/3.5	O-19-6°
Cream Buff	XXX 19''d	2.5 Y 8.0/6.5	OOY-17-6°
Cream Color	XVI 19'f	3.5 Y 8.5/5.5	OY-18-8°
Cream Color, measure of 1912		4.3 Y 8.2/3.7*	
Cream Color, measure of 1886		3.6 Y 7.3/3.6*	OY-18-8°
Cream Color, Palmer glossy		5.0 Y 8.1/5.5*	OY-18-8°
Cream Color, Palmer matte		5.0 Y 8.1/5.2*	OY-18-8°
Cream Color, Methuen		2.5 Y 8.5/4.0	
Cream Color, Guide swatch		3.5 Y 8.5/4.0*	

Citations for Cream Color

No useful citations for Cream Color were found in Bulletin 50.

Spectrum Yellow ~ Color 55

Spectrum Yellow is chosen to represent as pure a yellow as can be determined. It is a very conspicuous color of the spectrum where it has a much narrower range of wavelength than any other spectrum color. Methuen calls it spectrum yellow, with a wavelength from 568–580mμ and an 8.0Y 8.6/13.2 notation at 573mμ.

Palmer's Yellow swatch is nearly equivalent to the Spectrum Yellow of the NATURALIST'S COLOR GUIDE, perhaps with a slightly greater intensity. A spectrophotometer measured a 6.0Y 8.0/13.7 notation.

Ridgway's Lemon Yellow is very similar. It is shown in both *Color Standards* and *Nomenclature*. Notational measurements of his swatches differ from Hamly's but, strangely enough, nearly bracket it. Canary Yellow and Gamboge Yellow are shown only in *Nomenclature*. Both of them are used with some frequency in Bulletin 50, where they are often related to the more frequently used Lemon Yellow.

A lemon yellow swatch shown by Winsor & Newton is too pale to include in the color range, but their cadmium yellow matches very closely.

With so many color names to consider, all with nearly the same quality, it seems best simply to call them Spectrum Yellow.

The COLOR GUIDE swatch chosen for Spectrum Yellow has a 6.0Y 8.5/12.0 notation.

Notations of Colors Related to Spectrum Yellow

Color Name	Ridgway Notation	Hamly Notation	Villalobos Notation
Lemon Yellow	IV 23	7.5 Y 8.0/10.5	Y-18-12°
Lemon Yellow, measure of 1912		8.0 Y 7.7/ 9.2*	
Lemon Yellow, measure of 1886		6.8 Y 8.5/11.0*	
Pale Lemon Yellow	IV 23b	8.5 Y 8.5/10.0	Y-18-11°
Citron Yellow	XVI 23'b	7.5 Y 8.0/ 8.0	Y-17- 9°
Citron Yellow, measure of 1886		10.0 Y 8.4/ 7.7*	
Gamboge Yellow, measure of 1886		4.4 Y 8.0/11.9*?	Y-17-11°
Canary Yellow, measure of 1886		7.2 Y 8.5/ 9.0*	Y-17-10°
Picric Yellow	IV 23d	8.5 Y 8.5/ 9.0	Y-18-11°
Strontian Yellow	XVI 23'	7.5 Y 7.5/ 9.0	Y-16-11°
Spectrum Yellow, Methuen		8.0 Y 8.6/13.2	
Cadmium Yellow, Winsor & Newton		7.0 Y 8.7/13.0*	
Yellow, Palmer glossy		6.0 Y 8.0/13.7*	Y-18-12°
Yellow, Palmer matte		7.1 Y 8.0/11.5*	
Spectrum Yellow, Guide swatch		6.0 Y 8.5/12.0*	

140

Vol. VI   p. 332: face and throat bright lemon or pure gamboge yellow
     Keel-billed Toucan, *Ramphastos p. piscivorus*

Vol. IV   p. 675: below bright lemon or canary yellow
     Lictor Flycatcher, *Pitangus lictor*

Vol. IV   p. 672: below lemon or pure canary yellow
     Derby Flycatcher, *Pitangus sulphuratus derbianus*

Vol. IV   p. 665: below bright lemon or canary yellow; crownpatch pure yellow (canary or lemon)
     Boat-billed Flycatcher, *Megarynchus pitangus mexicanus*

Vol. IV   p. 662: below lemon or bright canary
     Golden-bellied Flycatcher, *Myiodynastes hemichrysus*

Vol. IV   p. 447: below lemon or bright canary
     Feraud's Flycatcher, *Myiozetetes t. texensis*

Vol. IV   p. 444: below lemon or bright canary
     Cayenne Flycatcher, *Myiozetetes c. cayanensis*

Vol. III   p. 306: outer rectrices between lemon and maize yellow
     Green Jay, *Xanthoura l. luxuosa*

Vol. III   p. 201: below lemon or canary yellow
     Golden Vireo, *Vireo h. hypochryseus*

Vol. III   p. 109: tail tips lemon yellow or chrome
     Cedar Waxwing, *Ampelis cedrorum*

Vol. II   p. 675: below light lemon or canary yellow
     ♂ Bryant's Yellow-throat, *Geothlypis rostrata*

*Vol. II   p. 673: rich lemon yellow
     ♂ Jalapa Yellow-throat, *Geothlypis trichas melanops*

Vol. II   p. 661: throat, chest lemon or canary yellow
     ♂ Maryland Yellow-throat, *Geothlypis t. trichas*

*Vol. II   p. 631: below clear lemon yellow
     ♂ Macgillivray's Warbler, *Oporornis tolmiei*

*Vol. II   p. 623: face (partly) and below clear lemon yellow
     Kentucky Warbler, *Oporornis formosa*

*Vol. II   p. 611: face (partly) and below clear lemon yellow
     ♂ Vitelline Warbler, *Dendroica vitellina*

Vol. II   p. 607: face (partly) and below gamboge or lemon yellow
     ♂ Prairie Warbler, *Dendroica discolor*

Vol. II   p. 604: face (partly) and below lemon, canary, or primrose
     ♂ Kirtland's Warbler, *Dendroica kirtlandii*

*Vol. II   p. 584: face, throat, chest lemon yellow
     Grace's Warbler, *Dendroica g. graciae*

Vol. II   p. 579: throat, chest lemon or gamboge yellow
     Yellow-throated Warbler, *Dendroica d. dominica*

*Vol. II   p. 565: face, neck rich lemon yellow
     ♂ Golden-cheeked Warbler, *Dendroica chrysoparia*

*Vol. II   p. 562: face, neck clear lemon yellow
     Black-throated Green Warbler, *Dendroica virens*

*Vol. II   p. 559: chest, breast clear lemon yellow
     Townsend's Warbler, *Dendroica townsendi*

*Vol. II   p. 546: crown patch, rump shades of lemon yellow (in spring)
     ♂ Myrtle Warbler, *Dendroica coronata*

*Vol. II   p. 533: rump lemon yellow; below rich lemon or gamboge yellow (plus streaks)

	♂ Magnolia Warbler, *Dendroica maculosa*
Vol. II	p. 529: below rich lemon or gamboge yellow
	♂ Bryant's Yellow Warbler, *Dendroica b. bryanti*
Vol. II	p. 527: below rich gamboge or lemon yellow
	♂ Panama Yellow Warbler, *Dendroica erythachorides*
Vol. II	pp. 515–526: misc. ♂ *Dendroica* subspecies, similarly rich lemon or gamboge
Vol. II	p. 508: head and below called only "rich yellow"
	♂ Yellow Warbler, *Dendroica a. aestiva*
*Vol. II	p. 456: below clear lemon yellow
	♂ Blue-winged Warbler, *Helminthophila pinus*
Vol. II	p. 442: head and below rich yellow, varying lemon yellow to cadmium yellow
	♂ Prothonotary Warbler, *Protonotaria citrea*
*Vol. II	p. 401: chest, breast lemon yellow
	Bahama Bananaquit, *Coereba bahamensis*
*Vol. II	p. 357: below bright lemon yellow
	Meadowlark, *Sturnella m. magna*
Vol. II	p. 308: rump and below (except chest) deep lemon yellow
	♂ Scott's Oriole, *Icterus parisorum*
*Vol. II	p. 305: some rectrices rich lemon yellow
	Yellow-tailed Oriole, *Icterus m. mesomelas*
*Vol. II	p. 191: rump, tail, coverts clear lemon yellow
	♂ Mexican Cacique, *Cassiculus melanicterus*
*Vol. II	p. 180: tail rich lemon yellow
	Montezuma Oropendola, *Gymnostinops montezuma*
*Vol. II	p. 124: below lemon yellow
	♂ White-throated Shrike-Tanager, *Lanio leucothorax*
Vol. II	p. 67: breast, upper belly bright yellow (lemon or gamboge)
	♂ Haitian Spindalis, *Spindalis multicolor*
Vol. II	p. 31: belly light yellow (lemon yellow to canary)
	♂ Jamaican Euphonia, *Pyrrhuphonia jamaica*
*Vol. II	p. 21: forehead area lemon yellow
	♂ Lesson's Euphonia, *Euphonia affinis*
Vol. II	p. 20: forehead and crown deep lemon or gamboge yellow
	♂ Yellow-crowned Euphonia, *Euphonia luteicapilla*
*Vol. II	p. 19: forehead and crown lemon yellow; underparts rich yellow (rich lemon, sometimes indian yellow)
	♂ Cabanis' Euphonia, *Euphonia gracilis*
Vol. II	p. 17: below rich yellow (deep lemon to indian yellow)
	♂ Tawny-capped Euphonia, *Euphonia anneae*
Vol. I	p. 621: back, rump, underparts bright yellow, varying from deep lemon yellow to light cadmium yellow
	♂ Vigor's Grosbeak, *Pheucticus chrysopeplus*
Vol. I	p. 463: face and below deep lemon or gamboge yellow
	White-naped Sparrow, *Atlapetes albinucha*
Vol. I	p. 118: below rich lemon yellow (*not* canary or citron yellow)
	♂ Central American Goldfinch, *Astragalinus psaltria croceus*
Vol. I	p. 109: general color pure lemon or canary yellow (in summer)
	♂ American Goldfinch, *Astragalinus t. tristis*
*Vol. I	p. 39: scapulars, rump clear lemon yellow
	♂ Evening Grosbeak, *Hesperiphona v. vespertina*

Citations for Canary Yellow (1886)

See citations for Lemon Yellow (Vol. IV, pp. 675, 672, 665, 662, 447, 444; Vol. III, p. 201; Vol. II, pp. 675, 661, 31; Vol. I, p. 109).

*Vol. VI	p. 347: below canary yellow (plus some bright red staining) Red-rumped Araçari, *Pteroglossus sanguineus*
*Vol. VI	p. 346: lower belly immaculate canary yellow Frantzius' Araçari, *Pteroglossus frantzii*
Vol. IV	p. 470: below canary or deep sulphur yellow Yellow Flycatcher, *Capsiempsis flaveola*
*Vol. IV	p. 367: below canary yellow Black-headed Tody-Flycatcher, *Todirostrum nigriceps*
Vol. IV	p. 334: below light canary or deep primrose Costa Rican Sharp-bill, *Oxyruncus eristatus frater*
Vol. III	p. 201: below lemon or canary yellow Golden Vireo, *Vireo h. hypochryseus*
*Vol. III	p. 163: face and below canary yellow Yellow-throated Vireo, *Lanivireo flavifrons*
Vol. III	p. 9: below between canary and naples yellow Alaskan Yellow Wagtail, *Budytes flavus alacensis*
*Vol. II	p. 628: below canary yellow Mourning Warbler, *Oporornis philadelphia*
Vol. II	p. 31: belly lemon to canary yellow Jamaican Euphonia, *Pyrrhuphonia jamaica*
Vol. II	p. 24: forehead area canary yellow or pale lemon Godman's Euphonia, *Euphonia godmani*
*Vol. I	p. 523: face and below canary yellow Mexican Yellow Finch, *Sicalis chrysops*
Vol. I	p. 114: below light yellow equal to canary yellow Arkansas Goldfinch, *Astragalinus p. psaltria*

Citations for Gamboge Yellow

See citations for Lemon Yellow (Vol. VI, p. 332; Vol. II, pp. 607, 579, 533, 529, 527, 526–515, 67, 20; Vol. I, p. 463).

*Vol. II	p. 452: below rich gamboge yellow Lawrence's Warbler, *Helminthophila lawrenceii*

Straw Yellow ~ Color 56

Straw Yellow is shown by Ridgway in both *Color Standards* and *Nomenclature*. It is also shown by Palmer (1962), where it has a slightly greenish cast, more so than Ridgway's swatches. However, there is a trend to greenish yellow in many straws.

Ridgway does not use the color very often in Bulletin 50. When he does, he usually relates it to other colors, ranging from canary yellow to sulphur and primrose yellow. It is probable that Straw Yellow is used more for descriptions of bills and legs (in fresh condition) than plumage.

The COLOR GUIDE swatch chosen for Straw Yellow has a 5.0Y 8.0/ 6.0 notation.

Notations of Colors Related to Straw Yellow

Color Name	Ridgway Notation	Hamly Notation	Villalobos Notation
Straw Yellow	XVI 21'd	6.0 Y 8.4/6.0	YYO-18-8°
Straw Yellow, measure of 1912		5.7 Y 8.2/5.0*	
Straw Yellow, measure of 1886		5.0 Y 8.2/5.4*	
Straw Yellow, Palmer glossy		7.0 Y 7.8/5.6*	YYO-18-8°
Straw Yellow, Palmer matte		7.0 Y 7.9/5.5*	YYO-18-8°
Straw Yellow, Methuen		5.5 Y 7.9/4.1	
Straw Yellow, Guide swatch		5.0 Y 8.0/6.0*	

Citations for Straw Yellow

Vol. VI   p. 282: below pale straw yellow or yellowish white (breast washed red) (♂ spring and summer)
Red-breasted Sapsucker, *Sphyrapicus r. ruber*

Vol. IV   p. 746: thighs straw or primrose yellow for ♂; greenish straw or olivaceous primrose yellow for ♀
Yellow-thighed Manakin, *Pipra m. mentalis*

Vol. IV   p. 642: below deep primrose or between straw and sulphur
Lawrence's Flycatcher, *Myiarchus l. lawrenceii*

Vol. IV   p. 629: below between pale canary and straw yellow, to deep primrose
Nutting's Flycatcher, *Myiarchus n. nuttingi*

Vol. IV   p. 614: below between straw and sulphur yellow
Crested Flycatcher, *Myiarchus crinitus*

Vol. IV   p. 577: below straw to primrose yellow
Western Flycatcher, *Empidonax d. difficilis*

Sulphur Yellow ~ Color 57

Sulphur Yellow is shown in *Color Standards* and *Nomenclature*. Somehow all the swatches depicting it, even the one shown by Methuen (1967), appear too greenish. However, that seems to be the consensus. The same considerations apply to Primrose Yellow, which is closely related to Sulphur Yellow.

Ridgway uses Sulphur Yellow frequently in Bulletin 50, where there are usually cross-references to other yellowish colors, such as canary, lemon, straw, primrose, even citron. More unequivocal citations are found for Primrose Yellow than for Sulphur Yellow, but the latter seems a better name to use.

The COLOR GUIDE swatch chosen for Sulphur Yellow has a 7.5Y 8.0/6.0 notation.

Notations of Colors Related to Sulphur Yellow

Color Name	Ridgway Notation	Hamly Notation	Villalobos Notation
Sulphur Yellow	V 25f	7.5 Y 8.4/5.0	YYL-19-9°
Sulphur Yellow, measure of 1912		0.7 GY 8.7/4.2*	
Sulphur Yellow, measure of 1886		9.4 Y 8.4/7.7*	YYL-19-9°
Sulphur Yellow, Methuen		1.0 GY 8.9/6.2	
Sulphur Yellow, Guide swatch		7.5 Y 8.0/6.0*	
Primrose Yellow	XXX 23''d	7.5 Y 8.2/6.5	Y-18-8°
Primrose Yellow, Methuen		1.0 GY 8.9/7.8	

Citations for Sulphur Yellow

Vol. VI p. 342: below canary or deep sulphur yellow
Collared Araçari, *Pteroglossus t. torquoitus*

Vol. VI p. 337: maxilla lemon yellow passing through sulphur yellow to light yellowish green
Swainson's Toucan, *Ramphastos swainsonii*

Vol. VI p. 309: below pale sulphur or primrose yellow
♂ Haitian Piculet, *Nesoctites micromegas*

Vol. VI p. 286: belly bright sulphur to nearly lemon yellow
♂ Williamson's Woodpecker, *Sphyrapicus thryoideus*

Vol. VI p. 274: middle underparts nearly sulphur to primrose yellow (in spring)
♂ Sapsucker, *Sphyrapicus v. varius*

147

Vol. III	p. 151: below mostly dull sulphur or primrose yellow
	Philadelphia Vireo, *Vireo philadelphica*
Vol. II	p. 61: vivid wing patch sulphur or light canary yellow
	Abbot Tanager, *Tanagra abbas*

Citations for Primrose Yellow

See citations for Sulphur Yellow (Vol. VI, pp. 309, 274, 102; Vol. IV, pp. 657, 582, 581, 465, 422, 405; Vol. III, p. 151) and Straw Yellow (Vol. IV, pp. 746, 642, 629, 577, 575, 550, 488, 417, 303; Vol. III, p. 188; Vol. II, p. 642).

Vol. VI	p. 287: belly primrose to nearly lemon yellow
	Williamson's Woodpecker, *Sphyrapicus thryoideus*
*Vol. IV	p. 652: below primrose (pale yellow)
	Sad Flycatcher, *Myiarchus barbirostris*
Vol. IV	p. 651: below deep primrose to light canary yellow
	Black-crested Flycatcher, *Myiarchus nigriceps*
*Vol. IV	p. 649: below primrose
	Olivaceous Flycatcher, *Myiarchus lawrenceii olivascens*
Vol. IV	p. 642: below deep primrose or between straw and sulphur yellow
	Lawrence's Flycatcher, *Myiarchus l. lawrenceii*
Vol. IV	p. 634: below pale primrose yellow
	Santo Domingo Flycatcher, *Myiarchus dominicensis*
*Vol. IV	p. 633: below primrose yellow
	Stolid Flycatcher, *Myiarchus stolidus*
Vol. IV	p. 632: below deep primrose yellow
	Yucatan Flycatcher, *Myiarchus yucatanensis*
*Vol. IV	p. 625: below primrose yellow
	Ash-throated Flycatcher, *Myiarchus c. cinerascens*
*Vol. IV	p. 621: below primrose yellow
	Mexican Crested Flycatcher, *Myiarchus m. mexicanus*
*Vol. IV	p. 617: below primrose yellow
	Ober's Flycatcher, *Myiarchus o. oberi*
*Vol. IV	p. 572: below primrose yellow
	Sierra Madre Flycatcher, *Empidonax pulverius*
Vol. IV	p. 493: below pale canary or deep primrose
	Salvin's Flycatcher, *Aphanotriccus capitalis*
Vol. IV	p. 490: rump and below primrose or light canary yellow
	Sulphur-rumped Myiobius, *Myiobius xanthopygus sulphureipygius*
*Vol. IV	p. 467: face and below primrose yellow (chest with olive tinge)
	Yellow-green Leptopogon, *Leptopogon flavovirens*
Vol. III	p. 210: below primrose or yellowish white
	Latimer's Vireo, *Vireo latimeri*

148

Yellow-Green ~ Color 58

Yellow-Green, a hyphenated color name, is shown in *Color Standards*. Ridgway also depicts many other colors in this color area, none of which are clearly used in Bulletin 50. There are no usable citations even for Yellow-Green. Instead of using these colors specifically, Ridgway resorts to generalized descriptive terms for pale greenish colors tinged strongly with yellow. It is possible that this situation is due to the fact that the colors are not shown in the earlier *Nomenclature*.

Palmer (1962) depicts a color called Yellow-Lime, which is nearly equivalent to Ridgway's Yellow-Green. Villalobos shows a color called lime, which is also a match to Yellow-Green. However, Palmer's Lime swatch does not equal that of Villalobos; it is substantially greener—somewhere between Ridgway's Apple Green and Emerald Green. These and several other colors in the range are shown in the tabulation.

The COLOR GUIDE swatch chosen for Yellow-Green has a 5.0GY 8.0/10.0 notation.

Notations of Colors Related to Yellow-Green

Color Name	Ridgway Notation	Hamly Notation	Villalobos Notation
Yellow-Green	VI 31	5.0 GY 8.0/10.0	L-16-10°
Yellow-Green, Hamly alternate		4.5 GY 7.2/11.0	L-16-10°
Yellow-Green, measure of 1912		2.6 GY 7.5/ 9.5*	
Yellow-Green, Guide swatch		5.0 GY 8.0/10.0*	
Clear Yellow-Green	VI 31b	3.5 GY 8.0/ 7.5	L-17-11°
Viridine Yellow	V 29b	1.0 GY 7.8/ 8.0?	L-17-12°
Yellow-lime, Palmer glossy		4.3 GY 7.6/11.2*	L-17-12°
Yellow-lime, Palmer matte		5.2 GY 7.6/10.5*	L-17-12°

Supplementary Table

Color Name	Ridgway Notation	Hamly Notation	Villalobos Notation
Lime, Palmer glossy		9.0 GY 6.7/12.2*	L-17-12°
Lime, Palmer matte		9.0 GY 6.5/12.2	

Citations for Yellow-Green

No useful citations for Yellow-Green were found in Bulletin 50.

Lime Green ~ Color 59

Lime Green is shown in *Color Standards* but not in *Nomenclature*. It is rarely used in Bulletin 50, and there is only one usable citation for it. Palmer (1962) shows a swatch called Lime-Green, but it does not match Ridgway's color; it is more greenish and much more intense. Limes, like olives, vary widely in color. Some are dark green, but turn nearly lemon yellow when ripe. Palmer shows three different colors containing the name of the fruit: Yellow-Lime, Lime (see Yellow-Green, Color 58), and Lime-Green. Ridgway shows only Lime Green, which Hamly gives a 1.0GY 6.5/5.0 notation. In the Villalobos designations Ridgway's swatch has a YYL-14-6° notation, whereas Palmer's registers GGL-12-12°. This is close to Ridgway's Emerald Green (GGL-14-12°) and approaches the Spectrum Green of the NATURALIST'S COLOR GUIDE.

The COLOR GUIDE swatch chosen for Lime Green has a 1.0GY 7.0/5.0 notation, which is close to Ridgway's Lime Green.

Notations of Colors Related to Lime Green

Color Name	Ridgway Notation	Hamly Notation	Villalobos Notation
Lime Green	XXXI 25″	1.0 GY 6.5/5.0	YYL-14-6°
Lime Green, measure of 1912		0.5 GY 6.6/5.5*	
Lime Green, Guide swatch		1.0 GY 7.0/5.0*	
Citron Green	XXXI 25″b	1.0 GY 7.0/5.0	YYL-15-7°
Citron Green, measure of 1912		1.3 GY 7.1/5.8*	
Deep Seafoam Green	XXXI 27″d	2.5 GY 8.0/5.0	YYL-17-7°
Pale Dull Green-Yellow	XVII 27′f	2.5 GY 8.2/4.0	YL-18-6°
Chrysolite Green	XXXI 27″b	2.5 GY 7.0/5.0	YL-15-6°
Pale Lumiere Green	XVII 29′f	2.5 GY 8.0/4.5	L-18-6°
Deep Chrysolite Green	XXXI 27″	2.5 GY 6.0/4.5	LLY-13-7°

Supplementary Table

Color Name	Ridgway Notation	Hamly Notation	Villalobos Notation
Lime-green, Palmer glossy		1.0 G 5.6/13.5*	GGL-12-12°*
Lime-green, Palmer matte		1.0 G 5.7/13.0*	

Citations for Lime Green

Vol. VIII   p. 9: remiges dull greenish yellow or yellowish green, nearest to lime green
Jacana, *Jacana s. spinosa*

Parrot Green ~ Color 60

Parrot Green is used frequently by Ridgway. This may be because he shows it in the early *Nomenclature* as well as *Color Standards,* or because it is so consistently the color of the plumage of parrots. Three other colors related to Parrot Green are shown in both color guides: Oil Green, Grass Green, and Chromium Green. These four colors produce more citations from Bulletin 50 than all the other greens combined. The colors have an admittedly distinctive quality, and perhaps I am being too bold in classifying them all in one "family."

The COLOR GUIDE swatch chosen for Parrot Green has a 5.0GY 5.5/5.5 notation, which is the same as that given by Hamly for Ridgway's color.

Notations of Colors Related to Parrot Green

Color Name	Ridgway Notation	Hamly Notation	Villalobos Notation
Parrot Green	VI 31k	5.0 GY 5.5/5.5	LLY- 8-5°
Parrot Green, measure of 1912		3.0 GY 5.0/4.4*	
Parrot Green, measure of 1886		6.0 GY 4.5/5.1*	
Parrot Green, Guide swatch		5.0 GY 5.5/5.5*	
Oil Green	V 27k	4.0 GY 5.0/4.5	LLY- 9-5°
Oil Green, measure of 1912		2.8 GY 4.8/4.9*	
Oil Green, measure of 1886		2.0 GY 6.3/6.1*	
Lettuce Green	V 29k	4.0 GY 5.8/4.5	LLY-10-6°
Spinach Green	V 29m	5.0 GY 4.6/4.0	L- 8-4°
Grass Green	VI 33k	7.5 GY 5.2/4.0	L- 8-4°
Grass Green, measure of 1912		6.0 GY 4.6/3.9*	
Chromium Green	XXXII 31"i	7.5 GY 5.5/5.0	L- 9-4°
Chromium Green, measure of 1912		6.2 GY 5.1/4.2*	
Cress Green	XXXI 29"k	5.0 GY 4.5/4.0	L- 7-3°
Forest Green	XVII 29m	7.5 GY 4.5/3.5	L- 7-4°

Citations for Parrot Green

*Vol. VII   p. 270: above general color parrot green (specimen 475328, central tail, measured 5.0GY 3.5/5.2 for *A. l. palmarum*) Cuban Parrot, *Amazona l. leucocephala*

Vol. VII   p. 266: nape, back bright parrot green Santo Domingo Parrot, *Amazona ventralis*

Vol. VII   p. 254: nape bright parrot green; back yellowish parrot green White-fronted Parrot, *Amazona a. albifrons*

152

Citations for Oil Green

See citations for Parrot Green (Vol. VII, pp. 179, 126; Vol. VI, p. 483; Vol. III, p. 306).

Vol. VII   p. 183: crown, nape clear paris green, then oil green, then again paris green on lower back and posterior scapulars
Tovi Paroquet, *Brotogeris jugularis*

Vol. VI   p. 466: back, rump (grayish green) bice to dull oil green
Chestnut-headed Motmot, *Momotus castaneiceps*

Vol. VI   p. 183: above (grayish yellowish green), nearly oil green
Cuban Green Woodpecker, *Xiphidiopicus p. percusius*

Citations for Grass Green

See citations for Parrot Green (Vol. VI, pp. 449, 448, 443).

*Vol. VI   p. 357: back, rump grass green
Blue-throated Toucanet, *Aulacorhynchus c. caeruleogularis*

*Vol. VI   p. 355: back, rump nearly pure grass green
Emerald Toucanet, *Aulacorhynchus p. prasinus*

Vol. VI   p. 354: back, rump yellowish grass green
Wagler's Toucanet, *Aulacorhynchus wagleri*

Vol. II   p. 43: above general color bright yellowish grass green
Blue-rumped Green Tanager, *Calospiza gyroloides*

153

Apple Green ~ Color 61

Apple Green is a little less grayish and more intense than Parrot Green, but the two colors are closely related. It is combined with other colors occasionally used by Ridgway—usually for smallish areas of plumage.

The COLOR GUIDE swatch for Apple Green has a 7.5GY 7.0/8.0 notation.

Notations of Colors Related to Apple Green

Color Name	Ridgway Notation	Hamly Notation	Villalobos Notation
Apple Green	XVII 29'	7.5 GY 6.5/ 8.0	L-15- 9°
Apple Green, measure of 1912		6.0 GY 6.6/ 7.0*	
Apple Green, Guide swatch		7.5 GY 7.0/ 8.0*	
Scheele's Green	VI 33i	7.5 GY 6.5/ 7.0	L-13-10°
Mineral Green	XVIII 31'	6.5 GY 6.5/ 8.0	L-14-11°
Viridine Green	VI 33d	6.5 GY 8.0/ 8.0	LLG-17-10°
Rivage Green	XVIII 31'b	6.5 GY 7.5/ 6.5	LLG-15- 9°
Veronese Green	XVIII 31'd	7.5 GY 7.5/ 6.0	LLG-16- 8°
Lumiere Green	XVII 29'b	7.5 GY 7.5/ 6.0	L-16- 8°
Vanderpoel's Green	VI 33b	7.5 GY 7.5/10.0	GGL-15-11°
Dull Green-Yellow	XVII 27'	5.0 GY 6.8/ 9.0	LLY-15- 9°

Citations for Apple Green

Vol. VII   p. 204: below (light yellowish green), bright apple green
                   Red-eared Parrot, *Pionopsitta h. haematotis*

Vol. VII   p. 192: side of head, below (light yellowish green), apple green
                   Mexican Parrotlet, *Psittacula c. cyanopygia*

Vol. VII   p. 160: below light apple green
                   Cuban Paroquet, *Aratinga euops*

Vol. VII   p. 159: below deep apple green
                   Socorro Paroquet, *Aratinga brevipes*

*Vol. VII   p. 153: below nearly apple green; above light apple green
                   Haitian Paroquet, *Aratinga c. chloroptera*

Vol. VII   p. 152: below, posteriorly, yellowish apple green
                   Finschi's Paroquet, *Aratinga finschi*

Vol. VII   p. 134: crown (not nape) bright apple green
                   Buffon's Macaw, *Ara ambigua*

Vol. VII   p. 126: posterior scapulars, mid-wing coverts, etc. (yellowish green), apple green to oil green
                   Red, Blue, and Green Macaw, *Ara chloroptera*

154

Vol. VI	p. 357: below (light yellowish green), between apple green and bice green
	Blue-throated Toucanet, *Aulacorhynchus c. caeruleogularis*
*Vol. III	p. 223: below (light yellowish green), apple green
	Green Shrike-Vireo, *Vireolanius p. pulchellus*
*Vol. II	p. 39: above general color (yellowish green), nearly apple green
	♂Emerald Tanager, *Calospiza f. florida*
*Vol. I	p. 589: pileum (bright yellowish green), apple green
	Leclancher's Nonpareil, *Cyanospiza leclancheri*
*Vol. I	p. 586: back (bright yellowish green or greenish yellow), apple green (back of specimen 365222 measured 0.5GY 4.9/7.2; back of 519404 measured 2.7GY 4.2/6.8)
	♂Painted Bunting, *Cyanospiza ciris*

Citations for Scheele's Green

Vol. VII	p. 262: head, neck (bright green), between scheele's and peacock
	Lesser Jamaican Parrot, *Amazona agilis*

155

Spectrum Green ~ Color 62

Spectrum Green is not a Ridgway color, although he does use Spectrum Red, Spectrum Blue, and Spectrum Violet. To judge from his descriptions, it would appear that Ridgway never found any plumage of a plain green color. Only two of his swatches come within the range of the COLOR GUIDE's Spectrum Green: Emerald Green and Vivid Green. There are only two usable citations in Bulletin 50 for the former and none for the latter.

Palmer shows a color called Green, and it is a completely acceptable color to represent Spectrum Green. In choosing the name, Ridgway's Emerald Green was considered, but it can be confused with Palmer's Emerald, which has a bluish quality. Vivid Green was a possible name. Its Villalobos notation (GE-11-11°) is very similar to that for Palmer's Green (G-10-12°). The Villalobos designation for Ridgway's Emerald Green is GGL-14-12°, which is visibly paler though still within the range of this color.

The COLOR GUIDE swatch chosen for Spectrum Green has a 2.5G 5.0/12.0 notation.

Notations of Colors Related to Spectrum Green

Color Name	Ridgway Notation	Hamly Notation	Villalobos Notation
Spectrum Green, Guide swatch		2.5 G 5.0/12.0*	
Vivid Green	VII 37	4.0 G 6.0/10.0	GE-11-11°
Vivid Green, measure of 1912		5.5 G 5.4/ 9.3*	
Green, Palmer glossy		2.1 G 4.8/14.4*	G-10-12°
Green, Palmer matte		2.8 G 4.9/12.8*	G-10-12°
Green, Methuen		1.0 G 6.3/10.8	
Green, Methuen		3.5 G 5.7/11.5	
Emerald Green	VI 35	2.0 G 6.7/10.0	GGL-14-12°
Emerald Green, measure of 1912		0.2 G 7.0/ 9.0*	
Emerald Green, measure of 1886		3.7 G 6.8/10.3*	
Cendre Green	VI 35b	1.0 G 7.5/ 8.0	G-13-11°
Lime-green, Palmer glossy		1.0 G 5.6/13.5*	GGL-12-12°*
Lime-green, Palmer matte		1.0 G 5.7/13.0*	

Supplementary Table

Emerald, Palmer glossy		0.7 BG 4.7/12.4*	E-10-12°
Emerald, Palmer matte		1.9 BG 5.3/11.0*	E-10-12°

Citations for Emerald Green

Vol. VII   p. 134: nape (bluish green), between emerald and malachite green

Buffon's Macaw, *Ara ambigua*

Vol. VI   p. 445: orbital area approaching emerald green

Narrow-billed Tody, *Todus angustirostris*

Paris Green ~ 63

Paris Green has been selected from a number of Ridgway colors of a fairly similar nature, including two tempting names: Malachite Green and Motmot Green. The Malachite Green swatch, in my opinion, shows no real likeness to the usual color of the mineral malachite. Motmot Green is an acceptable name, but it is not widely known. Paris Green is more familiar, and its color approximates that of the insecticide called paris green. Paris Green was therefore chosen as the name for this group of colors. Very few citations are available in Bulletin 50 to aid in an understanding of the relationships of the colors.

A supplementary tabulation shows colors not precisely related to Paris Green.

The COLOR GUIDE swatch chosen for Paris Green has a 2.5G 6.0/6.0 notation.

Notations of Colors Related to Paris Green

Color Name	Ridgway Notation	Hamly Notation	Villalobos Notation
Clear Fluorite Green	XXXII 33″b	1.0 G 7.0/5.0	LLG-14-6°
Malachite Green	XXXII 35″b	1.0 G 6.8/5.0	LG-13-5°
Malachite Green, measure of 1886		3.4 G 6.0/3.8*	LG-13-5°
Deep Malachite Green	XXXII 35″	1.0 G 5.6/5.0	G-12-6°
Fluorite Green	XXXII 33″	1.0 G 6.4/6.0	GGL-11-5°
Oriental Green	XVIII 33′	2.5 G 5.6/6.0	GGL-12-8°
Light Oriental Green	XVIII 33′b	2.5 G 6.6/6.0	GGL-14-7°
Motmot Green	XVIII 35′	2.5 G 5.8/7.0	G-11-8°
Motmot Green, measure of 1912		3.4 G 5.4/6.0*	G-11-8°
Paris Green	XVIII 35′b	2.5 G 6.8/6.5	G-13-8°
Paris Green, measure of 1912		2.5 G 6.0/5.5*	G-13-8°
Paris Green, Guide swatch		2.5 G 6.0/6.0*	

Supplementary Table

Color Name	Ridgway Notation	Hamly Notation	Villalobos Notation
Cobalt Green	XIX 37′b	3.5 G 7.0/ 5.0	GE-14-10°
Verdigris Green	XIX 37′	3.5 G 6.0/ 5.0	E-10- 8°
Viridian Green	VII 37′i	5.0 G 6.0/ 8.0	E- 8- 8°
Emerald, Palmer glossy		0.7 BG 4.7/12.4*	E-10-12°
Emerald, Palmer matte		1.9 BG 5.3/11.0*	E-10-12°

158

Citations for Malachite Green

*Vol. VII   p. 247: under the remiges (light dull bluish green), malachite
Double Yellow-head Parrot, *Amazona o. oratrix*

*Vol. VII   p. 234: under-remiges malachite green
Yellow-cheeked Parrot, *Amazona a. autumnalis*

Vol. VII   p. 232: under-remiges (dull bluish green), chromium green to
malachite green
Yellow-naped Parrot, *Amazona auropalliata*

*Vol. VII   p. 194: below (dull light bluish green), malachite green
♂Grayson's Parrotlet, *Psittacula insularis*

Vol. VII   p. 134: nape (bluish green), between emerald green and malachite
Buffon's Macaw, *Ara ambigua*

Citations for Paris Green

*Vol. VII   p. 183: crown, nape clear paris green, followed by oil green,
then on lower back and posterior scapulars by paris green
Tovi Paroquet, *Brotogeris jugularis*

Citations for Verdigris Green

Vol. VI   p. 472: breast, belly (dull bluish green), nearly verdigris green,
sometimes tinged with oil green
Lesser Broad-billed Motmot, *Electron platyrhynchus minor*

Vol. VI   p. 457: chin, throat, lower face (dull light bluish green), nearly
verdigris green
Lesson's Motmot, *Momotus l. lessoni*

Citations for Viridian Green

Vol. II   p. 383: largely except head glossy viridian green, becoming
more bluish in certain lights
♂Northern Green Honey Creeper, *Chlorophanes spiza guatemalensis*

Turquoise Green ~ 64
Turquoise Blue ~ 65

Turquoise Green is shown only in *Color Standards,* Turquoise Blue only in *Nomenclature.* The swatches clearly indicate a distinction between greenishness and bluishness. They can both be accepted as turquoise colors within the range of their respective "families." However, they are both somewhat too pale—and the green a little too green, the blue a little too blue—to match the gem mineral turquoise.

Two varieties of turquoise gemstones were studied, one from the southwestern United States and one from Europe. Only visual measurements were possible; the rounded shapes of the stones preclude the use of a spectrophotometer. A distinction was discernible between greenish and bluish qualities, but less than that shown by Ridgway. The stones from the United States were close to a 10.0BG 6.0/8.0 notation; those from Europe close to 2.5B 6.0/8.0. The former is used in the NATURALIST'S COLOR GUIDE for Turquoise Green, the latter for Turquoise Blue. They define the borderline between the two colors. Greener colors (approaching Ridgway's swatch) can still be called Turquoise Green and bluer colors Turquoise Blue.

Palmer (1962) shows two colors that clearly distinguish greenishness from bluishness: Turquoise and Turquoise-Cobalt. Spectrophotometric measurements showed that neither can be unequivocally included in the "family" of turquoise colors. They are too dark and too intense, both visually and notationally.

The COLOR GUIDE swatch chosen for Turquoise Green has a 10.0BG 6.0/8.0 notation; the swatch for Turquoise Blue has a 2.5B 6.0/8.0 notation.

Notations of Colors Related to Turquoise Green

Color Name	Ridgway Notation	Hamly Notation	Villalobos Notation
Turquoise Green	VII 41d	5.0 BG 7.0/6.0	ET-15-12°
Turquoise Green, measure of 1912		3.4 BG 7.2/5.4*	
Turquoise Green, U.S. gems		10.0 BG 6.0/8.0*	
Turquoise Green, Guide swatch		10.0 BG 6.0/8.0*	
Turquoise, Palmer glossy		9.7 BG 4.8/9.7*	T-11-12°
Turquoise, Palmer matte		9.2 BG 4.8/9.7*	
Venice Green	VII 41b	4.0 BG 6.0/8.0	ET-13-11°
Nile Blue	XIX 41'd	5.0 BG 7.0/6.0	T-14- 9°
Beryl Green	XIX 41'b	7.5 BG 6.4/8.0	T-13-10°

Notations of Colors Related to Turquoise Blue

Color Name	Ridgway Notation	Hamly Notation	Villalobos Notation
Turquoise Blue, measure of 1886		4.0 B 7.3/3.5*?	TC-13-10°
Turquoise Blue, European gems		2.5 B 6.0/6.0*	
Turquoise Blue, Guide swatch		2.5 B 6.0/6.0*	
Turquoise-cobalt, Palmer glossy		4.0 B 4.4/9.5*	TC- 9-12°*
Turquoise-cobalt, Palmer matte		5.0 B 4.5/9.0*	
Turquoise, Methuen		3.5 B 5.0/8.8	
Bremen Blue	XX 43'b	2.5 B 6.5/6.0	TC-13- 9°
Motmot Blue	XX 43'	2.5 B 5.2/6.5	TC-10-10°
Calamine Blue	VIII 43d	5.0 B 7.5/6.0	TC-15-11°

Citations for Turquoise
(Trending to Turquoise Green (1912))

Vol. VII	p. 254: crown (greenish blue), dull turquoise or light cerulean blue
	White-fronted Parrot, *Amazona a. albifrons*
Vol. VII	p. 134: greater wing-coverts (light greenish-blue), between turquoise and nile blue
	Buffon's Macaw, *Ara ambigua*
Vol. VII	p. 122: above general color turquoise to deep nile blue
	Blue and Yellow Macaw, *Ara ararauna*
Vol. II	p. 386: crown light turquoise or nile blue
	♂Blue Honeycreeper, *Cyanerpes cyaneus*
Vol. I	p. 584: head, neck, rump light cerulean or turquoise-blue, but changing to nile blue (light greenish blue) and back to duller blue
	♂Lazuli Bunting, *Cyanospiza amoena*

(Trending to Turquoise Blue (1886))

Vol. VII	p. 241: crown (conspicuously blue), turquoise to light azure blue
	Blue-crowned Parrot, *Amazona farinosa guatemalae*
*Vol. VII	p. 191: greater wing-coverts, lower back, rump bright turquoise blue
	Mexican Parrotlet, *Psittacula c. cyanopygia*
Vol. IV	p. 156: above rich turquoise or cerulean blue or sevres blue
	♂Mountain Bluebird, *Sialia arctica*
Vol. III	p. 315: back bright cerulean or deep turquoise blue; tail deeper, toward cobalt blue
	Yucatan Jay, *Cissilopha yucatanica*
*Vol. II	p. 46: belly turquoise blue
	♂Lavinia's Tanager, *Calospiza lavinia*
Vol. II	p. 43: rump and below cerulean or turquoise blue
	♂Blue-rumped Tanager, *Calospiza gyroloides*
Vol. I	p. 589: above light cerulean or deep turquoise blue (but back more or less tinged with green)
	♂Leclancher's Nonpareil, *Cyanospiza leclancheri*

161

Sky Blue ~ Color 66

Sky Blue is depicted in *Color Standards* but not in *Nomenclature*. It should be an easily determined color, although blue skies certainly vary in depth of hue. As shown by Ridgway and the NATURALIST'S COLOR GUIDE, it is moderately pale blue, and this paleness readily distinguishes it from Cerulean Blue.

Among possibly related colors, two should be discussed: Columbia Blue and Yale Blue, presumably so named because they were equivalent to the "school colors" of Columbia and Yale Universities. The Columbia Blue shown by Ridgway is close to that university's blue, but his Yale Blue varies so vastly from Yale's color that one wonders how it ever became one of his color names. Both Hamly and Villalobos give Yale Blue notations that would permit it to be included as a color related to Sky Blue, and visual comparisons confirm the similarity. But Yale University's blue measures close to Spectrum Blue, Cobalt, or Smalt Blue, though lacking the vivid quality of those colors. It is definitely not related to Sky Blue nor to Ridgway's Yale Blue swatch.

Columbia Blue contains a substantial amount of gray, which is indicated by Hamly's value notation of 5.5/ and by the Villalobos value of 6°. However, it is still acceptable as a related color.

The COLOR GUIDE swatch chosen for Sky Blue has a 2.5PB 7.0/7.0 notation, the same as that given by Hamly for Ridgway's swatch.

Notations of Colors Related to Sky Blue

Color Name	Ridgway Notation	Hamly Notation	Villalobos Notation
Sky Blue	XX 47'd	2.5 PB 7.0/7.0	CU-15-10°
Sky Blue, measure of 1912		2.5 PB 6.6/6.3*	
Sky Blue, Guide swatch		2.5 PB 7.0/7.0*	
Pale Methyl Blue	VIII 47d	10.0 B 7.2/6.0	CCU-15-10°
Light Squill Blue	XX 45'd	2.5 PB 6.8/6.0	CU-14-11°
Pale Cadet Blue	XXI 49'd	5.0 PB 6.5/5.0	UUC-14-11°
Columbia Blue	XXX 47''b	3.5 PB 5.5/8.0	UUC-12- 6°
Columbia Blue, measure of 1912		2.8 PB 5.6/4.0*	
Variance between Ridgway's Yale Blue and that of Yale University:			
Yale Blue	XX 47'b	3.5 PB 6.0/8.0	UUC-13-11°
Yale University Blue		6.5 PB 2.0/8.0*	UUC- 5-12°*

Citations for Sky Blue

No useful citations for Sky Blue were found in Bulletin 50.

Cerulean Blue ~ Color 67
Cobalt ~ Color 68

Cerulean Blue is used more often in Bulletin 50 than any other blue color. Ridgway shows it in both color guides. It is closely related to his Sevres Blue and Methyl Blue, although the latter shows a trend to Spectrum Blue. Cendre Blue is listed in the tabulation merely to record its notations; it is too pale a color to include in the "family."

Palmer's Cobalt is included in the NATURALIST'S COLOR GUIDE, although it is only slightly deeper than Cerulean Blue, based on visual and spectrophotometric comparisons of the swatches.

The COLOR GUIDE swatch chosen for Cerulean Blue has an 8.0B 5.0/10.0 notation, which is preferred to that given by Hamly for Ridgway's swatch. Cobalt has a 10.0B 3.8/11.0 notation.

Notations of Colors Related to Cerulean Blue and Cobalt

Color Name	Ridgway Notation	Hamly Notation	Villalobos Notation
Cerulean Blue	VIII 45	10.0 B 5.0/10.0	CCU- 8-12°
Cerulean Blue, measure of 1912		1.3 PB 4.7/10.5*	
Cerulean Blue, Guide swatch		8.0 B 5.0/10.0*	
Sevres Blue, 1886			CCU- 7-12°
Methyl Blue	VII 47	2.5 PB 5.0/14.0	CU- 8-13°
Cobalt, Palmer		10.0 B 3.8/10.5*	C- 8-12°
Cobalt, Guide swatch		10.0 B 3.8/11.0*	
Cendre Blue	VIII 43b	7.5 B 6.5/ 8.0	C-13-10°

Citations for Cerulean Blue

Vol. VII   p. 270: primaries, primary-coverts (light greenish blue) cendre blue to light cerulean blue (specimen 475328 (subspecies *palmarum*) measured 4.0PB 2.8/4.0 to 5.0PB 2.0/6.0—more purplish)
Cuban Parrot, *Amazona l. leucocephala*

Vol. VII   p. 254: crown (greenish blue) dull turquoise or light cerulean blue
White-fronted Parrot, *Amazona a. albifrons*

*Vol. VII   p. 194: lower back, rump, wing-coverts bright cerulean
Grayson's Parrotlet, *Psittacula insularis*

Vol. IV   p. 156: above turquoise, cerulean blue, or sevres blue
♂Mountain Bluebird, *Sialia arctica*

163

Vol. III	p. 315: back bright cerulean or deep turquoise blue; tail deeper toward cobalt blue
	Yucatan Jay, *Cissalopha yucatanica*
Vol. III	p. 313: lower back, rump cerulean or sevres blue
	San Blas Jay, *Cissalopha s. sanblasiana*
*Vol. III	p. 223: crown, nape cerulean blue
	Green Shrike-Vireo, *Vireolanius p. pulchellus*
Vol. II	p. 43: rump and below cerulean or turquoise blue
	♂Blue-rumped Tanager, *Calospiza gyroloides*
Vol. I	p. 590: above rich cerulean blue (more ultramarine or almost smalt on crown)
	♂Rosita's Bunting, *Cyanospiza rositae*
Vol. I	p. 589: above light cerulean or deep turquoise blue, etc.
	♂Leclancher's Nonpareil, *Cyanospiza leclancheri*
Vol. I	p. 584: head, neck, rump light cerulean or turquoise blue, etc.
	♂Lazuli Bunting, *Cyanospiza amoena*
Vol. I	p. 582: general color cerulean blue, changing to beryl green in some lights; head more ultramarine or french blue
	♂Indigo Bunting, *Cyanospiza cyanea*

Citations for Sevres Blue

See citations for Turquoise (Vol. IV, p. 156) and Cerulean Blue (Vol. IV, p. 156; Vol. III, p. 313).

Citations for Cendre Blue

See citations for Cerulean Blue (Vol. VII, p. 270).

164

Spectrum Blue ~ Color 69
Smalt Blue ~ Color 70

Spectrum Blue is shown in *Color Standards,* but it is rarely if ever used in Bulletin 50. Instead, Ridgway uses three other colors closely related to it: Cobalt Blue, Azure Blue, and Ultramarine Blue. The first two appear only in *Nomenclature,* the third in both *Nomenclature* and *Color Standards.* All of these colors present a somewhat confusing picture; in Bulletin 50 the colors are related to different colors: azure sometimes to turquoise, sometimes to cobalt; cobalt to cerulean; ultramarine to cerulean and smalt.

To complicate the situation further, Palmer (1962) shows swatches for Cobalt and Ultramarine; they match acceptably well with the notations indicated for them by Villalobos. Measured by spectrophotometer, Cobalt registered near Cerulean and Ultramarine near Smalt Blue. In other words, Palmer's Cobalt is a slightly paler blue than the COLOR GUIDE's Spectrum Blue, and his Ultramarine more violaceous, closer to Ridgway's Smalt Blue.

In order to clarify the colors, those with the same or similar names published by Methuen and by Winsor & Newton were examined. Both of their cobalt blues and Methuen's ultramarine blue measure within the range of Spectrum Blue. Winsor & Newton's french ultramarine compares well with the COLOR GUIDE's Smalt Blue range.

The COLOR GUIDE swatch chosen for Spectrum Blue has a 5.5PB 4.0/14.0 notation; Smalt Blue has an 8.0PB 3.5/12.0 notation.

Notations of Colors Related to Smalt Blue

Color Name	Ridgway Notation	Hamly Notation	Villalobos Notation
Smalt Blue	IX 53i	8.5 PB 3.2/14.0	U-6-12°
Smalt Blue, measure of 1912		7.7 PB 3.5/11.6*	
Smalt Blue, measure of 1886		8.0 PB 4.1/12.5*	
Smalt Blue, Guide swatch		8.0 PB 3.5/12.0*	
Phenyl Blue	IX 53	7.5 PB 3.5/16.0	U-6-13°
Hay's Blue	IX 53k	7.5 PB 3.2/10.5	U-5-11°
Helvetia Blue	IX 51k	7.5 PB 3.2/10.0	U-5-10°
Ultramarine, Palmer glossy		7.5 PB 2.0/12.0*	U-5-12°
French Ultramarine, Winsor & Newton		7.9 PB 3.1/17.0*	

Notations of Colors Related to Spectrum Blue

Color Name	Ridgway Notation	Hamly Notation	Villalobos Notation
Spectrum Blue	IX 49	5.0 PB 4.0/16.0	UUC-8-13°
Spectrum Blue, measure of 1912		6.0 PB 4.0/14.0*	
Spectrum Blue, Guide swatch		5.5 PB 4.0/14.0*	
Cobalt Blue, 1886			CU/UUC-8-13°
Cobalt Blue, measure of 1886		5.0 PB 4.3/10.5*	
Cobalt Blue, Methuen		5.0 PB 4.3/13.4	
Cobalt Blue, Winsor & Newton		5.7 PB 4.6/12.4*	
Azure Blue, 1886			CU/UUC-8-13°
Bradley's Blue	IX 51	6.0 PB 4.2/16.0	UUC-7-13°
Ultramarine Blue	IX 49i	6.5 PB 3.5/16.0	UUC-7-13°
Ultramarine Blue, measure of 1912		6.2 PB 3.7/13.4*	
Ultramarine Blue, measure of 1886		6.3 PB 3.9/14.5*	
Ultramarine Blue, Methuen		5.5 PB 3.7/12.2	

Citations for Cobalt Blue (1886)

Vol. VII p. 128: primary coverts, remiges dull cobalt blue (specimen 474215 measured 6.2PB 2.6/4.8 and 5.0PB 2.0/4.0 to 2.0/6.0 —more purplish)
Red, Blue, and Yellow Macaw, *Ara Macao*

Vol. III p. 347: greater wing-coverts, secondaries, rectrices rich cobalt or azure blue (broadly white-tipped)
Blue Jay, *Cyanocitta c. cristata*

*Vol. III p. 344: nearly entire plumage dull cobalt blue
Unicolored Jay, *Aphelocoma u. unicolor*

*Vol. III p. 327: pileum, nape, side neck dull cobalt blue
California Jay, *Aphelocoma c. californica*

Vol. III p. 315: tail deeper than cerulean, toward cobalt blue
Yucatan Jay, *Cissilopha yucatanica*

Citations for Azure Blue (1886)

Vol. VII p. 241: crown (conspicuously blue), turquoise to light azure blue
Blue-crowned Parrot, *Amazona farinosa guatemalae*

*Vol. VII p. 128: rump azure blue (specimen 474215 measured 5.8PB 4.0/5.0 to 5.0PB 5.0/8.0—more purplish)
Red, Blue, and Yellow Macaw, *Ara Macao*

*Vol. III p. 361: rump and below azure blue
Aztec Jay, *Cyanocitta stelleri azteca*

Vol. III	p. 347: greater wing-coverts, secondaries, rectrices rich cobalt or azure blue (broadly white-tipped)
	Blue Jay, *Cyanocitta c. cristata*
*Vol. III	p. 327: wings, tail dull azure blue
	California Jay, *Cyanocitta c. californica*
*Vol. III	p. 326: pileum, nape, scapulars, wings, tail dull azure blue
	Florida Jay, *Aphelocoma cyanea*

Citations for Ultramarine Blue

Vol. IV	p. 148: rump, tail (less violaceous than smalt), more ultramarine (in spring)
	♂Mexican Bluebird, *Sialia m. mexicana*
Vol. IV	p. 142: above average hue in spring between smalt and ultramarine blue
	♂Bluebird, *Sialia s. sialis*
Vol. I	p. 607: general color dull ultramarine blue (in spring)
	♂Blue Grosbeak, *Guiraca c. caerulea*
Vol. I	p. 590: crown more ultramarine than cerulean back, almost smalt blue
	♂Rosita's Bunting, *Cyanospiza rositae*
Vol. I	p. 582: head more ultramarine or french blue than general cerulean color
	♂Indigo Bunting, *Cyanospiza cyanea*

Citations for Smalt Blue

*Vol. IV	p. 148: above rich smalt blue (in spring)
	♂Mexican Bluebird, *Sialia m. mexicana*
Vol. IV	p. 142: above (bright blue), average hue in spring between smalt and ultramarine blue
	♂Bluebird, *Sialia s. sialis*
*Vol. II	p. 386: below, etc., smalt blue
	♂Blue Honeycreeper, *Cyanerpes cyaneus*
Vol. I	p. 590: crown toward ultramarine, almost smalt blue
	♂Rosita's Bunting, *Cyanospiza rositae*
Vol. I	p. 586: head, neck (purplish blue), smalt or hyacinth blue (specimen 365222 measured 8.5PB 3.0/4.7; 519404 measured 8.3PB 2.4/3.8 and 7.5PB 2.0/4.0 to 2.0/6.0)
	♂Painted Bunting, *Cyanospiza ciris*

Campanula ~ Color 71

Campanula is used by Ridgway in Bulletin 50 about as frequently as most other bluish colors, and it is shown in both of his color guides. Many colors similar to Campanula are shown in the tabulation.

The COLOR GUIDE swatch chosen for Campanula has a 9.0PB 6.0/ 10.0 notation, the same as that given by Hamly for Ridgway's swatch.

Notations of Colors Related to Campanula

Color Name	Ridgway Notation	Hamly Notation	Villalobos Notation
Campanula Blue	XXIV 55*b	9.0 PB 6.0/10.0	U-13-11°
Campanula Blue, measure of 1912		7.5 PB 5.7/ 7.9*	
Campanula Blue, measure of 1886		7.5 PB 5.8/ 7.9*	
Campanula, Guide swatch		9.0 PB 6.0/10.0*	
Light Soft Blue-Violet	XXIII 55'b	9.0 PB 6.0/10.0	U-13-12° (to UUV)
Wistaria Blue	XXIII 57'b	9.0 PB 5.8/10.0	UUV-13-11°
Wistaria Blue, measure of 1912		8.7 PB 5.5/ 9.2*	
Deep Chicory Blue	XXIV 57*b	10.0 PB 5.4/10.0	UUV-12-10°
Deep Lavender Blue	XXI 53'b	8.5 PB 5.8/10.0	U-13-11°
Light Grayish Violet-Blue	XXIV 53*b	9.5 PB 5.8/10.0	U-13-11°
Lavender Blue	XXI 53'd	8.5 PB 6.5/ 9.0	U-14-11°
Lavender Blue, measure of 1912		7.4 PB 6.4/ 7.5*	
Deep Wedgwood Blue	XXI 51'd	7.5 PB 7.0/10.0	U-14-11°
Flax-flower Blue	XXI 51'b	7.5 PB 6.0/11.0	U-13-11°

Citations for Campanula Blue

Vol. VII   p. 229: forehead, chin, throat (violaceous blue), nearly campanula
♂Bonquet's Parrot, *Amazona arausiaca*

Vol. VII   p. 227: forehead, face (light violaceous blue), between azure blue and campanula
♂Santa Lucia Parrot, *Amazona versicolor*

Vol. III   p. 347: most upper parts (grayish violet-blue), dull campanula blue
Blue Jay, *Cyanocitta c. cristata*

*Vol. III   p. 306: crown, nape campanula blue
Green Jay, *Xanthoura l. luxuosa*

*Vol. II   p. 60: head, neck (dull purplish), campanula blue
♂Abbot Tanager, *Tanagra abbas*

Vol. I   p. 591: fore-crown, lower nape, rump (light purplish blue or bluish purple), mauve to campanula or flax-flower blue (in summer)
♂Varied Bunting, *Cyanospiza versicolor*

Spectrum Violet ~ Color 72

Spectrum Violet is one of Ridgway's spectrum color names. However, no citations are found in Bulletin 50 to confirm his use of the color. Neither are there useful citations for any of the other colors of this group, except for the darker Hyacinth Blue. Spectrum Violet is similar to Palmer's Ultramarine Violet.

The COLOR GUIDE swatch chosen for Spectrum Violet has a 10.0PB 3.5/14.0 notation.

Notations of Colors Related to Spectrum Violet

Color Name	Ridgway Notation	Hamly Notation	Villalobos Notation
Spectrum Violet	X 59	10.0 PB 3.5/14.0	UV-6-12°
Spectrum Violet, measure of 1912		0.24 P 3.2/14.7*	
Spectrum Violet, Guide swatch		10.0 PB 3.5/14.0*	
Royal Purple	X 59i	1.0 P 3.5/ 8.0	UV-5-11°
Hyacinth Blue	X 55k	9.0 PB 2.6/ 8.0	UUV-4-10°
Deep Blue-Violet	X 55i	9.0 PB 2.8/14.0	UUV-5-11°
Blue-Violet	X 55	9.0 PB 3.5/16.0	UUV-6-12°
Bluish Violet	X 57	9.0 PB 3.8/16.0	UUV-5-12°
Violet Ultramarine	X 57i	9.0 PB 3.4/12.0	UUV-4-12°
Ultramarine-violet, Palmer		0.2 P 2.0/15.0*	UUV-5-12°
Violet, Methuen		2.5 P 3.6/17.2	
Violet Blue, Methuen		9.0 PB 3.1/17.7	
Bluish Violet, Methuen		1.0 P 3.3/16.1	
Royal Purple, Methuen		3.0 P 2.7/11.8	

Citations for Hyacinth Blue

Vol. VII   p. 254: some primaries (deep blue), nearly hyacinth blue
            White-fronted Parrot, *Amazona a. albifrons*
Vol. I   p. 586: head, neck (purplish blue) smalt or hyacinth blue
            ♂Painted Bunting, *Cyanospiza ciris*

Indigo Blue ~ Color 73
Cyanine Blue ~ Color 74

Indigo Blue and Cyanine Blue are used occasionally in Bulletin 50. While they are both similar and dark, they cannot be called closely related. Ridgway shows Indigo Blue as including much gray and Cyanine Blue as having a richer hue.

The COLOR GUIDE swatch chosen for Indigo Blue has a 5.0PB 3.0/3.0 notation; Cyanine Blue has a 7.5PB 3.0/6.0 notation. These notations are the same as those given by Hamly for Ridgway's swatches.

Notations of Colors Related to Indigo Blue

Color Name	Ridgway Notation	Hamly Notation	Villalobos Notation
Indigo Blue	XXXIV 47″m	5.0 PB 3.0/3.0	U-5-6°
Indigo Blue, Guide swatch		5.0 PB 3.0/3.0*	
Dark Tyrian Blue	XXXIV 47″k	5.0 PB 3.8/3.0	U-6-6°
Acetin Blue	XXXV 49″k	7.5 PB 3.5/4.0	U-5-6°
Alizarine Blue	XXI 57′m	6.0 PB 3.5/6.0	U-5-7°

Notations of Colors Related to Cyanine Blue

Color Name	Ridgway Notation	Hamly Notation	Villalobos Notation
Cyanine Blue	IX 57m	7.5 PB 3.0/6.0	U-4-10°
Cyanine Blue, Guide swatch		7.5 PB 3.0/6.0*	
Berlin Blue	VIII 47m	5.0 PB 2.6/5.0	UUC-4- 9°
Azurite Blue	IX 53m	8.5 PB 2.5/6.0	UUV-4- 9°
Prussian Blue	IX 49m	3.5 PB 3.2/9.0	UUC-4-10°

Citations for Indigo Blue

Vol. II	p. 34: above largely dull indigo blue
	♂Cherrie's Tanager, *Buthraupis caeruleogularis*
Vol. II	p. 33: above dull grayish indigo blue
	♂Arce's Tanager, *Buthraupis arcaei*

Citations for Cyanine Blue

Vol. I	p. 601: general color deep cyanine or marine blue (but facial areas, rump, etc., bright cobalt or azure blue)
	♂Blue Bunting, *Cyanocompsa p. parellina*
Vol. I	p. 599: general color dull berlin blue to almost cyanine
	♂Guiana Blue Grosbeak, *Cyanocompsa cyanoides*

Mauve ~ Color 75
Lilac ~ Color 76
Lavender ~ Color 77

Mauve is rarely, if ever, used by Ridgway, although he shows it in both *Color Standards* and *Nomenclature*. It is seldom used ornithologically, but it is a common popular color name. It is decidedly purplish in hue. Lilac is similar, but is much paler and is close to Ridgway's Light Mauve. Lavender is another related color with a popular name.

The COLOR GUIDE swatch chosen for Mauve has a 3.5P 5.5/10.0 notation; Lilac has a 5.0P 7.0/5.0 notation; Lavender has a 3.0P 7.0/3.0 notation.

Notations of Colors Related to Mauve

Color Name	Ridgway Notation	Hamly Notation	Villalobos Notation
Mauve	XXV 63'b	2.5 P 6.0/ 6.0	VVU-12-10°
Mauve, measure of 1912		3.5 P 5.4/ 9.5*	
Mauve, measure of 1886		5.0 P 5.8/ 8.0*	
Mauve, Guide swatch		3.5 P 5.5/10.0*	
Lavender-Violet	XXV 61'b	2.5 P 6.0/10.0	UV-13-11°
Chinese Violet	XXV 65'b	5.0 P 5.4/ 8.0	V-12-11°
Amparo Purple	XI 63b	5.0 P 5.5/10.0	V-11-11°
Hortense Violet	XI 61b	5.0 P 6.0/12.0	VVU-12-12°
Light Lavender-Violet	XXV 61'd	1.5 P 6.0/10.0	UV-14-11°
Light Hortense Violet	XI 61d	5.0 P 6.0/10.0	V-13-12°

Notations of Colors Related to Lavender

Color Name	Ridgway Notation	Hamly Notation	Villalobos Notation
Lavender	XXXVI 59''F	2.5 P 7.6/5.0	VVU-16-5°
Lavender, measure of 1912		5.6 P 7.2/2.8*	
Lavender, measure of 1886		3.0 P 8.1/2.6*	
Deep Lavender	XXXVI 59''d	2.5 P 6.5/6.0	UV-13-8°
Deep Lavender, measure of 1912		3.5 P 6.1/4.4*	
Lavender, Guide swatch		3.0 P 7.0/3.0*	

Notations of Colors Related to Lilac

Color Name	Ridgway Notation	Hamly Notation	Villalobos Notation
Lilac	XXV 65'd	5.0 P 7.5/5.0	V-15-10°
Lilac, measure of 1912		5.4 P 6.6/6.6*	
Lilac, measure of 1886		6.5 P 7.0/4.8*	
Light Mauve	XXV 63'd	2.5 P 7.0/8.0	VVU-15-10°
Lilac, Guide swatch		5.0 P 7.0/5.0*	

Citations for Mauve, Lavender & Lilac

No useful citations for Mauve, Lavender, or Lilac were found in Bulletin 50.

Plumbeous ~ Color 78

Plumbeous is a medium gray color with a clearly visible bluish cast. It is defined as a color resembling the metal lead. Ridgway shows it in both color guides and uses it with some frequency in Bulletin 50. In the Plumbeous Kite, *Ictinea plumbea,* the so-called plumbeous areas are too mixed with other colors to permit dependable measurements. The Belted Kingfisher measures fairly close to Plumbeous, although it is overly bluish and dark. The NATURALIST'S COLOR GUIDE swatch has been chosen somewhat arbitrarily as nearest to the color of lead.

The COLOR GUIDE swatch chosen for Plumbeous has a 5.0PB 4.0/1.5 notation.

Notations of Color Names Related to Plumbeous

Color Name	Ridgway Notation	Hamly Notation	Villalobos Notation
Plumbeous	LII 49''''''b	7.5 PB 5.6/1.5	UUV-12-12°
Deep Plumbeous	LII 49''''''	7.5 PB 5.0/1.5	UUV-10-12°
Plumbeous, measure of 1912		7.5 PB 5.8/1.3*	
Plumbeous, measure of 1886		6.6 PB 6.1/1.2*	
Specimen 361964 as measured		6.0 B 3.4/0.5*	
Specimen 374018 as measured		5.0 B 3.1/0.2*	
Plumbeous, Guide swatch		5.0 PB 4.0/1.5*	

Citations for Plumbeous & Related Colors

Vol. XI	p. 125: miscellaneous small areas plumbeous black ♂Plumbeous Kite, *Ictinea plumbea*
Vol. XI	p. 102: above dark plumbeous or plumbeous black (gray phase) ♂Hook-billed Kite, *Chondrohierax u. uncinatus*
*Vol. XI	p. 12: neck ruff plumbeous King Vulture, *Sarcorhamphus papa*
Vol. VI	p. 415: above bluish gray (nearly plumbeous) (specimen 361964 measured 6.0B 3.4/0.5) Belted Kingfisher, *Streptoceryle a. alcyon*
Vol. VI	p. 409: above bluish plumbeous Ringed Kingfisher, *Streptoceryle t. torquata*
Vol. VI	p. 398: breast slate gray or deep plumbeous ♂Marraganti Nun Bird, *Monasa minor*
Vol. IV	p. 838: above gray, between plumbeous and slate-gray ♂Cinereous Becard, *Pachyrhamphus cinereus*

Vol. IV	p. 829: below slate-gray or plumbeous gray
	♂Gray-bellied Becard, *Pachyrhamphus polychropterus cinerei-ventris*
Vol. III	p. 456: above deep bluish gray or plumbeous
	Pygmy Nuthatch, *Sitta p. pygmacea*
Vol. III	p. 453: above bluish gray or plumbeous (specimen 374018 measured 5.0B 3.1/0.2)
	Brown-headed Nuthatch, *Sitta pusilla*
Vol. III	p. 450: above bluish gray or deep plumbeous
	Red-breasted Nuthatch, *Sitta canadensis*
Vol. III	p. 441: above bluish gray (between slate-gray or No. 6 and plumbeous)
	White-breasted Nuthatch, *Sitta c. carolinensis*
Vol. III	p. 237: above light bluish gray (between No. 8 and cinereous)
	♂Northern Shrike, *Lanius borealis*
Vol. II	p. 699: above bluish gray or deep plumbeous
	♂Du Bus' Red-breasted Chat, *Granatellus venustus*

174

Glaucous ~ Colors 79 & 80

Glaucous may be a pale yellowish green color with a very weak intensity or chroma, if judged by Ridgway's swatch in *Color Standards*. That is also the dictionary's definition of the color. Hamly gives Ridgway's swatch a 7.5GY 8.0/2.0 notation, which may be correct, but that color is too rich to accept as Glaucous.

Ridgway depicts about twenty other swatches of a glaucous nature with a variety of hues: yellowish, greenish, bluish, and grayish. One wonders what quality relates them all to glaucous. No citations were found in Bulletin 50 to clarify the question. However, two large gulls use the name: the Glaucous-winged Gull and the Glaucous Gull. They were measured to see if their plumage had a glaucous quality.

Visual comparisons showed that the Glaucous-winged Gull, the darker of the two very pale birds, very nearly matched a neutral gray with a notation of N 5.0. The Glaucous Gull matched a notation of about N 6.0 to N 6.5. The specimens could also be matched visually to swatches with notations of 5.0Y 5.0/1.0 and 5.0Y 6.0/1.0, respectively.

Spectrophotometric measurements registered notations of 2.8Y 5.4/0.4 and 3.3Y 6.4/0.7, respectively. It should be remembered that when notations of such low chromas are involved (/1.0,/0.4,/0.7), the hues (5.0Y, 2.8Y, 3.3Y) cannot be unequivocally accepted as accurate, although some slight hue is properly indicated. These plumages are best described as of approximately N 5.0 and N 6.0 neutral gray. Ridgway appears to have arrived at a similar conclusion. He describes the Glaucous-winged Gull as "pale neutral gray" (close to N 6.0) and the Glaucous Gull as "pale gull gray" (nearer to N 8.0).

Despite the inconclusive evidence as to the precise hue of the color Glaucous, two swatches are depicted in the NATURALIST'S COLOR GUIDE. Perhaps they could be slightly greener, but not as colorful as Ridgway's Glaucous swatch.

The COLOR GUIDE swatches chosen for Glaucous have notations of 5.0Y 5.0/1.0 and 6.0Y 6.0/1.0.

Notations of Colors Related to Glaucous

Color Name	Ridgway Notation	Hamly Notation		Villalobos Notation
Glaucous	XLI 29'''f	7.5 GY	8.0/2.0	LG-16-4°
Glaucous, measure of 1912		8.9 GY	7.6/1.9*	
Glaucous-Gray	XLIII 37''''f	2.5 B	7.5/1.0	CU-14-7°
Glaucous-Gray, measure of 1912		4.6 B	6.4/0.6*	
Glaucous-winged Gull, spectrophotometer		2.8 Y	5.4/0.4*	
Glaucous-winged Gull, visual measure		N 5.0/neutral gray* and 5.0 Y	5.0/1.0*	
Glaucous-Gull, spectrophotometer		3.3 Y	6.4/0.7*	
Glaucous Gull, visual measure		N 6.0/neutral gray* and 5.0 Y	6.0/1.0*	
Glaucous, Guide swatch		5.0 Y	5.0/1.0* and	
		6.0 Y	6.0/1.0*	

Citations for Glaucous

No useful citations for Glaucous were found in Bulletin 50 except for the two discussed in the notes.

Pearl Gray ~ Color 81

Pearl Gray is shown in both *Color Standards* and *Nomenclature*. It is also shown by Palmer (1962). The name, of course, derives from pearls. Natural pearls vary materially in hue, sometimes faintly bluish and sometimes slightly greenish yellow in tone. They are usually close to some neutral gray, however, perhaps with a notation of N 7.5.

Spectrophotometric measurements showed consistently similar values (about 7.5/) and low chromas. But hue notations varied widely and should be ignored. At very weak or low chromaticities, it really means very little whether instrument measurements measure bluish, yellowish, or greenish hues.

The COLOR GUIDE swatch chosen for Pearl Gray has a 5.0GY 8.0/1.0 notation, which is close to that of Palmer's swatch. Some color may be visible, but it may seem bluish, yellowish, or greenish to different eyes.

Notations of Colors Related to Pearl Gray

Color Name	Ridgway Notation	Hamly Notation	Villalobos Notation
Pearl Gray	LII 35'''''f	2.5 B 8.0/0.2	CCU-16-11°
Pearl Gray, Palmer		5.7 GY 7.5/1.1*	
Pearl Gray, Methuen		N 7.5/neutral gray	
Specimen 494660		1.0 GY 7.1/0.7*	
Pearl Gray, Guide swatch		5.0 GY 8.0/1.0*	
Pearl Gray, measure of 1912		5.0 Y 8.6/0.8*	

Citations for Pearl Gray

Vol. IV   p. 863: above pale bluish gray, nearest No. 8, or between lavender and pearl gray
Fraser's Erator, *Erator albitorques*

Vol. IV   p. 789: above white, shaded with pale bluish gray (pearl gray) (specimen 494660 measured 1.0GY 7.1/0.7)
Antonia's Cotinga, *Carpodectes antoniae*

Vol. IV   p. 788: above pale gray (paler than pearl gray)
Snowy Cotinga, *Carpodectes nitidus*

Blackish Neutral Gray ~ Color 82
Dark Neutral Gray ~ Color 83
Medium Neutral Gray ~ Color 84
Light Neutral Gray ~ Color 85
Pale Neutral Gray ~ Color 86

Grayish colors present a very special problem. They must be thought of in two distinctly different categories. If they include a substantial degree of spectral hue, they are indeed colors and are then entitled to a color name. If they are devoid of spectral hue, depicting only darkness or lightness, ranging from nearly black to nearly white, they are called neutral grays. Neutral grays are preferably designated by numbers rather than names. However, the NATURALIST'S COLOR GUIDE follows Palmer's (1962) use of descriptive color terminology, correlating his names with the appropriate Munsell numerical designations.

Some grayish colors that include spectral hue have been discussed previously (Dusky Brown, Dark Grayish Brown, Fuscous, Drab, Olive, Olive-Gray, Grayish Olive, Smoke Gray, Plumbeous, Glaucous, and Pearl Gray), and some of these approach the neutral gray category.

Ridgway uses both categories of gray in Bulletin 50. *Color Standards* identifies "purer" grays by numbers, ranging from No. 1 for black to No. 10 for nearly white. Unfortunately, *Color Standards* distinguishes between neutral grays (depicting a range of grayishness) and carbon grays (the percentage of lampblack and chinese white used to produce degrees of lightness or darkness in his color swatches). Ridgway gives names as well as numbers to his grays, so it is not clear whether or not some spectral hue is present. For example, his Pale Neutral Gray swatch appears to match his No. 8 swatch, which he calls Gull Gray. His descriptive words, notational number, and color name seem to be synonymous for all practical purposes, and all these grays are practically devoid of spectral hue.

The Munsell Color Co. publishes a series of some twenty neutral grays, identified with numerical notations—N 0.5 (nearly black) to N 9.5 (nearly white). Villalobos also shows twenty neutral grays, numbered from 0 (black) to 20 (white). Methuen shows only six neutral gray swatches. Palmer (1962) also shows six neutral gray swatches, identified by descriptive terms only, such as pale gray, dark gray, etc.

The tabulation correlates the notations, names, and descriptive terms of all of these authorities for the neutral grays. It should be regarded with some reservations, though, because exact equivalency of tonal quality is not to be found.

The COLOR GUIDE swatches chosen for the neutral grays are: Blackish Neutral Gray N 2.5, Dark Neutral Gray N 4.0, Medium Neutral Gray N 5.0, Light Neutral Gray N 6.0, and Pale Neutral Gray N 7.5.

Citations for Neutral Gray

Many citations for gray are found in Bulletin 50, but there is some confusion as to whether neutral grays or hued grays are intended. The following citations are presumed to be of a neutral character because of the terminology used. Only a few of the great many such citations are included, merely to quote Ridgway's usage. They indicate the need for scientific measurements.

Vol. IX	p. 120: upper breast gull gray or deep gull gray
	Guiana Wood-Rail, *Aramides c. cajanea*
Vol. VIII	p. 752: above slate color or deep slate-gray
	Xantus' Murrelet, *Endomychura hypoleuca*
Vol. VIII	p. 649: back, wings pale bluish gray (nearly No. 8, gull gray)
	Little Gull, *Hydrocoloeus minutus*
Vol. VIII	p. 623: back, wings pale gray (between No. 8 and pale, neutral gray)
	Ring-billed Gull, *Larus delawarensis*
Vol. VIII	p. 612: back, wings pale gray (between pale and pallid neutral gray)
	Herring Gull, *Larus argentatus*
Vol. VIII	p. 597: back, wings pale neutral gray (specimen 358091 measured 2.8Y 5.4/0.4)
	Glaucous-winged Gull, *Larus glaucescens*
Vol. VIII	p. 585: back, wings pale gull gray (specimen 442693 measured 3.3Y 6.4/0.7)
	Glaucous Gull, *Larus hyperboreus*
Vol. VIII	p. 479: above pale gray (between gull gray and pallid neutral gray)
	Gull-billed Tern, *Gelochelidon nilotica*
Vol. VII	p. 456: crown gull gray or light gull gray
	♂Cassin's Dove, *Leptotila c. cassini*
Vol. IV	p. 837: breast, etc., light gray (about No. 7)
	♂Arce's Becard, *Pachyrhamphus a. albo-griseus*
Vol. IV	p. 789: below pale gray (nearly No. 8)
	♀Antonia's Cotinga, *Carpodectes antoniae*
Vol. IV	p. 788: below pale gray (about No. 8)
	♂Snowy Cotinga, *Carpodectes nitidus*
Vol. IV	p. 706: above gray (about No. 6)
	Gray Kingbird, *Tyrannus d. dominicensis*
Vol. IV	p. 642: chest light gray (about No. 8 or 9)
	Lawrence's Flycatcher, *Myiarchus l. lawrenceii*

Vol. IV	p. 632: chest light gray (about No. 8)
	Yucatan Flycatcher, *Myiarchus yucatanensis*
Vol. IV	p. 629: chest pale gray (about No. 10)
	Nutting's Flycatcher, *Myiarchus n. nuttingi*
Vol. IV	p. 625: chest much paler gray (about No. 10)
	Ash-throated Flycatcher, *Myiarchus c. cinerascens*
Vol. IV	p. 621: chest pale gray (about No. 10)
	Mexican Crested Flycatcher, *Myiarchus m. mexicanus*
Vol. IV	p. 617: chest light gray (about No. 8 or 9)
	Ober's Flycatcher, *Myiarchus o. oberi*
Vol. IV	p. 614: chest gray (about No. 8)
	Crested Flycatcher, *Myiarchus crinitus*
Vol. IV	p. 179: breast ash gray (No. 6 or 7)
	St. Vincent Solitaire, *Myadestes sibilans*

Neutral Gray Correlations

Munsell Notation	Palmer Name	Villalobos Notation	Ridgway Notation and Name	Methuen Notation and Name
N 9.5 near white	white	20 white	— White	1A1 white, chalky, snowy white
N 8.0	pale gray	17	10 Pale Gull Gray	1B1 pale gray, grayish white (pearl white)
N 7.5		16		
N 7.0		15		1C1 light gray, pastel gray (pearl gray)
N 6.5	light gray	14	9 Light Gull Gray, Pallid Neutral Gray	
N 6.0		13	8 Gull Gray, Pale Neutral Gray	1D1 platinum (steel gray, lead gray)
N 5.5		12 to 11	7 Deep Gull Gray, Light Neutral Gray	
N 5.0	medium gray	10 to 9	6 Dark Gull Gray, Neutral Gray	1E1 medium gray, neutral gray (pewter)
N 4.0	dark gray	7	5 Slate-Gray, Deep Neutral Gray	
N 3.5 to 3.0		6 to 5	4 Slate Color, Dark Neutral Gray	1F1 dark gray
N 2.5 to 2.0	blackish gray	4	3 Blackish Slate, Dusky Neutral Gray	
N 1.5 to 1.0	black	3, 2, 1	2 Slate-Black	G1 blackish gray
N 0.5		0	1 Black	H1 grayish black

REFERENCES

Artists Water Colours, Wealdstone Harrow, Middlesex, England: Winsor & Newton, Ltd.

ASTM Standard Method D-1535, Philadelphia: American Society for Testing & Materials, 1968.

Bent, A. C., *Life History of North American Birds,* U.S. National Museum Bulletins, New York: Dover Publications, 20 Volumes, 1924–1958.

Blake, Emmet R., *Birds of Mexico,* Chicago: University of Chicago Press, 1953.

Chapanis, A., "Color Names for Color Space," *American Scientist,* Vol. 53, 1965, pp. 327–346.

Chapman, Frank M., *Handbook of North American Birds,* New York: D. Appleton & Co., 1921.

Color Cascade, Baltimore: Munsell Color Co., 1970.

Color Harmony Manual, Chicago: Center for Advanced Research and Design, Container Corporation of America, Third Edition, 1948.

Eisenmann, Eugene, *The Species of Middle American Birds,* Transactions of the Linnaean Society of New York, Vol. VII, 1955.

Evans, R. M., *An Introduction to Color,* New York: John Wiley & Sons, 1948.

Foss, C. E., D. Nickerson, and W. C. Granville, "Analysis of the Ostwald Color System," *Journal of the Optical Society of America,* Vol. 34, No. 7, 1944, pp. 361–381.

Granville, Walter, "Colorimetric Data for Third Edition *Color Harmony Manual,*" Libertyville, Illinois: Granville Color Service, 1948.

Hamly, D. H., "The Ridgway Color Standards with a Munsell Notation Key," *Journal of the Optical Society of America,* Vol. 39, No. 7, 1949, pp. 592–599.

Hausman, L. A., *The Illustrated Encyclopedia of American Birds,* Garden City, New York: Garden City Publishing Co., 1947.

Imhof, Thomas A., *Alabama Birds,* University of Alabama Press, 1962.

Kelly, K. L., K. S. Gibson, and D. Nickerson, "Tristimulus Specification of the *Munsell Book of Color* from Spectrophotometric Measurements," and following authors and articles, *Journal of the Optical Society of America,* Vol. 33, No. 7, 1943, pp. 355–418.

———— and J. B. Judd, *The ISCC-NBSS Method of Designating Colors and a Dictionary of Color Names,* NBSS Circular 553, Washington, DC: Government Printing Office, 1955. (Supplement: *ISCC-NBSS Color-Name Charts Illustrated with Centroid Colors,* Standard Sample 2106, Washington, DC: National Bureau of Standards, 1965).

Kornerup, A., and J. H. Wanscher, *Methuen Handbook of Color,* London: Methuen & Co., Ltd., Second Edition, 1967.

Land, Hugh C., *Birds of Guatemala,* Wynnewood, Pennsylvania: Livingston Publishing Co., 1970.

Lanyon, Wesley E., "The Middle American Populations of the Crested Flycatcher *Myiarchus tyrannulus,*" *Condor,* Vol. 62, No. 5, 1960, pp. 341–350.

_____ "Specific Limits and Distribution of Ash-throated and Nutting Fly-catchers," *Condor*, Vol. 63, No. 6, 1961, pp. 421–449.

Maerz, A., and M. Rea Paul, *A Dictionary of Color*, New York: McGraw Hill Book Co., Second Edition, 1950.

Norris, R. B., "Colors of Stomach Linings of Certain Passerines," *Wilson Bulletin*, Vol. 73, No. 4, 1961, pp. 380–383.

Palmer, Ralph S., *Handbook of North American Birds*, New Haven: Yale University Press, Vol. 1, 1962.

_____ "Color Specification," *A New Dictionary of Birds*, A. L. Thomson, editor, New York: McGraw Hill Book Co., 1964, p. 144.

_____ and E. M. Reilly, Jr., *A Concise Color Standard*, Albany: American Ornithologists Union Handbook Fund, 1956.

Peterson, Roger Tory, Bird Guide Series, Boston: Houghton Mifflin Co.

Phillips, Allan R., "Geographic Variation in *Empidonax traillii*," *The Auk*, Vol. 65, No. 4, 1948, pp. 507–514.

_____ "An Ornithological Comedy of Errors. . . ," *The Auk*, Vol. 86, No. 4, 1969, pp. 605–623.

_____ and Wesley E. Lanyon, "Additional Notes on the Flycatchers of Eastern North America," *Bird Banding*, Vol. 41, No. 3, 1970, pp. 190–197.

_____ and Wesley E. Lanyon, and M. A. Howe, "Identification of the Flycatchers of Eastern North America. . . ," *Bird Banding*, Vol. 37, No. 3, 1966, pp. 153–171.

Pickford, Kenneth D., "Colour Definition," *Ibis*, Vol. 112, No. 1, 1970, p. 117.

Ridgway, Robert, *A Nomenclature of Colors . . . for Ornithologists*, Boston: Little Brown & Co., 1886.

_____ *Color Standards and Color Nomenclature*, Washington, DC, 1912.

_____ and H. Friedmann, *The Birds of North and Middle America*, Bulletin 50, Washington, DC: U.S. National Museum, Smithsonian Institution, Eleven Volumes, 1901–1950.

Selander, R. K., and R. F. Johnston, "Evolution in the House Sparrow. . . ," *Condor*, Vol. 69, No. 3, 1967, pp. 217–258.

Smithe, Frank B., *The Birds of Tikal*, Garden City, New York: Natural History Press, 1966.

Villalobos-Dominguez, C., and J. Villalobos, *Atlas de los Colores*, Buenos Aires: Libreria El Ateneo Editorial, 1947.

Webster, J. D., "A Revision of the Rose-Throated Becard," *Condor*, Vol. 65, No. 5, 1963, pp. 383–399.

Willis, Edwin O., "The Behavior of Plain-brown Woodcreepers, *Dendrocindra fuliginosa*," *Wilson Bulletin*, Vol. 84, No. 4, 1972, pp. 377–420.

Zimmer, J. T., Letter, *Science*, Vol. 108, No. 2805, 1948, p. 356.

APPENDIX A
CORRELATED TABLES

Appendix A combines three important lists into a single tabulation. First are Ridgway's color names and notations derived from *Color Standards*. The names of the colors are listed alphabetically, followed by the symbols he gives them.

Next is a list of Hamly's notations for Ridgway's color swatches, made from visual measurements and shown in Munsell notations. The third list is from the conversion tables of Villalobos, showing the Ridgway colors in terms of the Villalobos notation system.

The Correlated Notes of the NATURALIST'S COLOR GUIDE SUP-PLEMENT have relied heavily on the three lists, which are now all out of print. Although there are several disagreements among the various notations, so many measurements agree that publishing them in a unified tabulation seems desirable.

A color name in boldface type indicates that the color is represented by a swatch in the NATURALIST'S COLOR GUIDE. An asterisk indicates that the color is discussed in the Correlated Notes, although not shown in the guide.

Ridgway Color Name	Ridgway Notation		Hamly Notation	Villalobos Notation
Absinthe Green	XXXI	29''	5. GY 6.6/ 5.5	LLY-12- 7°
Acajou Red	XIII	1'i	2.5 R 4.0/ 5.5	RS- 8- 6°
*Acetin Blue	XXXV	49''k	7.5 PB 3.5/ 4.0	U- 5- 6°
Ackermann's Green	XVIII	35'k	3.5 G 4.5/ 4.0	G- 9- 5°
Aconite Violet	XXXVII	63''	7.5 P 4.5/ 7.0	VVM- 8- 6°
Ageratum Violet	XXXVII	63''b	7.5 P 5.4/ 6.0	V-10- 7°
Alice Blue	XXXIV	45''b	10. B 6.5/ 4.0	UUC-12- 6°
*Alizarine Blue	XXI	51'm	6. PB 3.5/ 6.0	U- 5- 7°
Alizarine Pink	XIII	1'd	2.5 R 6.5/ 7.5	MR-14-12°
Amaranth Pink	XII	69d	2.5 RP 6.0/11.0	R- 6-12°
Amaranth Purple	XII	69i	8.5 RP 3.2/10.0	RS-14- 8°
Amber Brown	III	13k	5. YR 3.8/ 8.0	OOS- 8-10°
Amber Yellow	XVI	21'b	5. Y 8.0/ 8.0	YYO-17- 9°
American Green	XLI	33'''i	10. GY 5.2/ 3.0	GGL-10- 3°
Amethyst Violet	XI	61	3.5 P 5.0/14.0	
			or	VVU- 6-12°
			2.5 P 3.4/15.0	
Amparo Blue	IX	51b	7.5 PB 5.8/12.0	UUC-10-12°
*Amparo Purple	XI	63b	5. P 5.5/10.0	V-11-11°
Andover Green	XLVII	25''''i	5. GY 4.8/ 2.0	Y- 6- 2°
Aniline Black	L	69''''m	7.5 P 2.5/ 1.0	M- 4- 2°
Aniline Lilac	XXXV	53''d	10. PB 7.0/ 8.0	UUV-14- 6°
Aniline Yellow	IV	19i	2.5 Y 6.0/ 9.0	OOY-14-10°
Anthracene Green	VII	39m	5. BG 3.5/ 4.5	ET- 6- 5°
Anthracene Purple	XLIV	69'''k	5. RP 2.6/ 3.0	M- 6- 3°
Anthracene Violet	XXV	61'k	2.5 P 3.5/ 6.0	VVU- 6- 6°
Antimony Yellow	XV	17'b	1. YR 7.4/ 7.5	OOY-16- 8°
Antique Brown	III	17k	7.5 YR 4.4/ 5.0	O- 8- 9°
Antique Green	VI	35m	2.5 G 3.5/ 2.0	G- 7- 3°
Antwerp Blue	VIII	45k	10. B 3.0/ 5.0	CCU- 5-12°
Apple Green	XVII	29'	7.5 GY 6.5/ 8.0	L-15- 9°
Apricot Buff	XIV	11'b	5. YR 7.0/ 7.5	OOS-14- 8°
Apricot Orange	XIV	11'	2.5 YR 6.0/ 9.0	SO-13-10°
*Apricot Yellow	IV	19b	2.5 Y 7.8/ 9.5	OOY-18-12°
*Argus Brown	III	13m	6. YR 3.8/ 4.0	OOS- 6- 8°
Argyle Purple	XXXVII	65''b	2.5 RP 5.2/ 8.0	VM-10- 5°
*Army Brown	XL	13'''i	5. YR 4.4/ 2.5	SO- 9- 3°
Artemisia Green	XLVII	33''''	2.5 BG 5.6/ 1.0	GE-10- 2°
Asphodel Green	XLI	29'''	5. GY 5.8/ 3.5	L-11- 4°
Aster Purple	XII	67i	1. P 3.4/10.0	MR- 6- 8°
*Auburn	II	11m	2.5 YR 3.4/ 4.0	SO- 5- 5°
Auricula Purple	XXVI	69'k	3.5 RP 3.2/ 8.0	MMV- 6- 6°
*Avellaneous	XL	17'''b	10. YR 7.0/ 3.0	OOS-13- 4°
*Azurite Blue	IX	53m	8.5 PB 2.5/ 6.0	UUV- 4- 9°
Barium Yellow	XVI	23'd	7.5 Y 8.0/ 7.5	Y-18-10°
Baryta Yellow	IV	21f	5. Y 8.4/ 6.0	YYO-19-12°
*Bay	II	7m	10. R 2.6/ 6.0	SSO- 5- 6°
*Begonia Rose	I	1b	5. R 5.6/11.5	RS-13-12°
Benzo Brown	XLVI	13''''i	5. YR 4.5/ 2.0	SO- 9- 2°

186

Ridgway Color Name	Ridgway Notation		Hamly Notation	Villalobos Notation
Benzol Green	VII	41	7.5 BG 5.0/ 8.0	TC- 9-11°
			or	
			8. BG 5.2/10.5	
*Berlin Blue	VIII	47m	5. PB 2.6/ 5.0	UUC- 4- 9°
Beryl Blue	VIII	43f	5. B 8.0/ 6.0	TC-17-10°
*Beryl Green	XIX	41'b	7.5 BG 6.4/ 8.0	T-13-10°
Bice Green	XVII	29'k	7.5 GY 5.0/ 4.0	L- 9- 3°
Biscay Green	XVII	27'i	6. GY 5.2/ 4.5	LLY-10/11-4°
Bishop's Purple	XXXVII	65''	2.5 RP 4.0/ 7.2	VM- 8- 6°
*Bister	XXIX	15''m	9. YR 3.5/ 2.0	OOS- 5- 5°
Bittersweet Orange	II	9b	1. YR 6.0/10.0	SO-15-11°
Bittersweet Pink	II	9d	2.5 YR 7.0/ 8.5	SO-16-11°
*Black	LIII	NG	N 2.2/	N-2/
*Blackish Brown (1)	XLV	1''''m	10. RP 1.0/ 3.0	MR- 4- 2°
*Blackish Brown (2)	XLV	5''''m	2.5 R 2.8/ 1.0	S- 4- 2°
*Blackish Brown (3)	XLV	9''''m	5. YR 2.5/ 1.0	S- 4- 2°
Blackish Green-Blue	VIII	43m	5. B 3.2/ 5.0	C- 4- 2°
Blackish Green-Gray	LII	35'''''m	5. G 2.4/ 0.2	UUC- 3- 2°
Blackish Mouse Gray	LI	15'''''m	7.5 YR 3.0/ 0.5	VVM- 4- 1°
Blackish Plumbeous	LII	49'''''k	7.5 PB 3.0/ 1.0	U- 4- 2°
Blackish Purple	XI	65m	10. P 2.0/ 3.5	VM- 3- 8°
Blackish Red-Purple	XII	67m	5. RP 2.4/ 5.0	MR- 3- 6°
Blackish Slate	LIII	CGm	N 2.6/	UUV- 3- 1°
Blackish Violet	X	59m	1. P 2.5/ 5.0	UV- 3- 9°
Blackish Violet-Gray	LII	59'''''m	2.5 P 3.2/ 0.5	UV- 2- 2°
Blanc's Blue	XX	47'k	2.5 PB 4.0/ 7.0	UUC- 6-10°
Blanc's Violet	XXIII	59'k	1.5 P 3.5/ 3.0	UV- 6- 7°
Blue-Violet	X	55	9. PB 3.5/16.0	UUV- 6-12°
Blue-Violet Black	XLIX	57''''m	2.5 P 3.0/ 0.5	UV- 2- 2°
Bluish Black	XLIX	49''''m	6.5 PB 2.0/ 1.5	U- 3- 4°
Bluish Glaucous	XLII	37'''f	7.5 BG 8.0/ 2.5	ET-16- 4°
Bluish Gray-Green	XLII	41'''	2.5 BG 5.5/ 2.0	ET-10- 4°
Bluish Lavender	XXXVI	57''d	2.5 P 6.2/ 6.0	UV-13- 8°
Bluish Slate-Black	XLVIII	45''''m	5. PB 2.5/ 2.0	UUC- 3- 3°
*Bluish Violet	X	57	9. PB 3.8/16.0	UUV- 5-12°
*Bone Brown	XL	13'''m	5. YR 2.6/ 1.5	S- 5- 2°
Bordeaux	XII	71k	5. RP 3.4/ 6.0	R- 4- 7°
Bottle Green	XIX	37'm	3.5 G 3.4/ 3.0	E- 4- 4°
*Bradley's Blue	IX	51	6. PB 4.2/16.0	UUC- 7-13°
Bradley's Violet	XXIII	59'	2.5 P 4.5/ 9.0	UV- 8- 9°
Brazil Red	I	5i	7.5 R 4.0/10.5	S- 7-11°
*Bremen Blue	XX	53'b	2.5 B 6.5/ 6.0	TC-13- 9°
*Brick Red	XIII	5'k	10. R 3.6/ 5.0	SSO- 6- 5°
*Bright Chalcedony Yellow	XVII	25'	2.5 GY 8.0/ 7.0	YL-16-10°
*Bright Green-Yellow	V	27	4. GY 7.0/12.0	LLY-17-12°
Brownish Drab	XLV	9'''''	5. YR 4.6/ 2.0	S-10- 3°
Brownish Olive	XXX	19''m	2.5 Y 3.8/ 3.0	O- 5- 2°
Brownish Vinaceous	XXXIX	5'''b	5. R 6.0/ 3.0	S-12- 6°

187

Ridgway Color Name	Ridgway Notation		Hamly Notation	Villalobos Notation
*Brussels Brown	III	15m	7.5 YR 3.5/ 4.0	OOS- 6- 7°
Buckthorn Brown	XV	17'i	10. YR 5.0/ 7.0	O-11- 9°
Buff-Pink	XXVIII	11''d	5. YR 7.0/ 5.0	SO-16- 7°
*Buffy Brown	XL	17'''i	10. YR 5.5/ 4.0	O- 9- 3°
Buffy Citrine	XVI	19'k	2.5 Y 5.0/ 5.0	OY- 9- 6°
*Buffy Olive	XXX	21''k	5. Y 5.0/ 4.0	OOY- 9- 2°
Buff-Yellow	IV	19d	2.5 Y 8.0/ 7.0	OOY-18-10°
Burn Blue	XXXIV	47''f	5. PB 7.5/ 7.0	U-15- 6°
Burnt Lake	XII	71m	2.5 RP 3.4/ 4.0	RS- 4- 6°
Burnt Sienna	II	9k	10. R 3.4/ 8.0	SO- 5-12°
Burnt Umber	XXVIII	9''m	2.5 YR 3.0/ 2.0	SSO- 5- 3°
*Cacao Brown	XXVIII	9''i	2.5 YR 4.6/ 4.0	SO-10- 5°
Cadet Blue	XXI	49'i	6. PB 4.5/ 9.0	U- 9-11°
Cadet Gray	XLII	45'''b	2.5 PB 6.5/ 3.0	UUC-12- 5°
*Cadmium Orange	III	13	5. YR 6.0/14.0	OOS-14-12°
*Cadmium Yellow	III	17	7. YR 6.6/12.0	OOY-16-12°
			or	
			0.5 Y 7.2/13.0	
*Calamine Blue	VIII	43d	5. B 7.5/ 6.0	TC-15-11°
Calla Green	V	25m	1. GY 4.0/ 4.5	YL- 7- 4°
Calliste Green	VI	31i	5. GY 6.5/ 8.0	LLY-14-11°
*Cameo Brown	XXVIII	7''k	2.5 YR 4.0/ 3.0	SSO- 7- 4°
Cameo Pink	XXVI	71'f	5. RP 8.0/ 9.0	MR-17- 8°
Campanula Blue	XXIV	55*b	9. PB 6.0/10.0	U-13-10°
Capri Blue	XX	43'i	2.5 B 4.4/ 5.5	TC- 8- 8°
Capucine Buff	III	13f	9. YR 8.0/ 6.0	OOS-18-10°
Capucine Orange	III	13d	6. YR 7.0/ 8.5	O-17-11°
Capucine Yellow	III	15b	8.5 YR 7.0/11.0	O-17-12°
Carmine	I	1i	5. R 3.4/12.5	RS- 7-12°
Carnelian Red	XIV	7'	10. R 6.0/11.5	SO-13-10°
*Carob Brown	XIV	9'm	2.5 YR 2.0/ 4.0	SSO- 4- 4°
Carrot Red	XIV	7'b	10. R 8.5/ 6.0	SO-14-10°
*Cartridge Buff	XXX	19''f	2.5 Y 8.5/ 3.5	O-19- 6°
Castor Gray	LII	35'''''i	5. G 4.8/ 0.5	T- 8- 1°
Cedar Green	VI	31m	5. GY 4.5/ 4.0	L- 6- 3°
Celandine Green	XLVII	33''''b	2.5 BG 6.5/ 1.0	G-11- 2°
Cendre Blue	VIII	43b	7.5 B 6.5/ 8.0	C-13-10°
*Cendre Green	VI	35b	1. G 7.5/ 8.0	G-13-11°
Cerro Green	V	27m	5. GY 4.2/ 3.5	LLY- 7- 2°
Cerulean Blue	VIII	45	10. B 5.0/10.0	CCU- 8-12°
			or	
			9.5 B 4.7/11.0	
*Chaetura Black	XLVI	17''''m	10. YR 2.4/ 1.0	SSO- 4- 1°
*Chaetura Drab	XLVI	17''''k	10. YR 3.6/ 1.0	OOS- 6- 3°
*Chalcedony Yellow	XVII	25'b	10. Y 8.2/ 6.5	YYL-17- 8°
Chamois	XXX	19''b	2.5 Y 7.5/ 6.0	OOY-15- 6°
Chapman's Blue	XXII	49*i	6.5 PB 4.5/ 8.0	UUC- 9- 9°
*Chartreuse Yellow	XXXI	25''d	10. Y 8.2/ 6.0	YYL-18- 8°
Chatenay Pink	XIII	3'f	6. R 7.5/ 6.0	S-16- 8°

188

Ridgway Color Name	Ridgway Notation		Hamly Notation	Villalobos Notation
Chessylite Blue	XX	45'k	9. B 3.5/ 5.0	CU- 7- 9°
Chestnut	II	9m	10. R 3.0/ 5.0	SSO- 5- 6°
*Chestnut-Brown	XIV	11'm	2.5 YR 3.4/ 3.0	SO- 5- 3°
Chicory Blue	XXIV	57*d	9. PB 6.4/ 8.0	UUV-14-10°
China Blue	XX	45'i	10. B 5.0/ 5.5	CU- 9- 9°
*Chinese Violet	XXV	65'b	5. P 5.4/ 8.0	V-12-11°
*Chocolate	XXVIII	7''m	2.5 YR 2.6/ 2.5	S- 5- 4°
*Chromium Green	XXXII	31''i	7.5 GY 5.5/ 5.0	L- 9- 4°
*Chrysolite Green	XXXI	27''b	2.5 GY 7.0/ 5.0	YL-15- 6°
Chrysopraise Green	VII	37b	10. GY 7.0/ 6.5	G-14-11°
Cinereous	LII	49'''''d	7.5 PB 6.5/ 1.5	UV-14- 1°
Cinnamon	XXIX	15''	7.5 YR 6.2/ 6.0	OOS-13- 6°
Cinnamon-Brown	XV	15'k	7.5 YR 3.8/ 4.0	OOS- 7- 6°
*Cinnamon-Buff	XXIX	17''b	1. Y 7.2/ 6.0	O-15- 7°
*Cinnamon-Drab	XLVI	13''''	5. YR 5.5/ 2.5	SO-11- 3°
Cinnamon-Rufous	XIV	11'i	2.5 YR 5.0/ 8.0	OOS-10- 8°
Citrine	IV	21k	5. Y 4.5/ 5.0	YYO- 9- 8°
*Citrine-Drab	XL	21'''i	5. Y 5.2/ 4.0	OY- 8- 4°
*Citron Green	XXXI	25''b	10. Y 7.0/ 5.0	YYL-15- 7°
*Citron Yellow	XVI	23'b	7.5 Y 8.0/ 8.0	Y-17- 9°
Civette Green	XVIII	31'k	9. GY 4.8/ 4.0	LG- 8- 3°
Claret Brown	I	5m	7.5 R 3.0/ 5.0	S- 4- 7°
Clay Color	XXIX	17''	1. Y 5.4/ 5.0	O-13- 7°
Clear Cadet Blue	XXI	49'	6. PB 5.5/10.0	UUC-10-11°
*Clear Dull Green Yellow	XVII	27'b	5. GY 7.5/ 7.0	YL-17-10°
*Clear Fluorite Green	XXXII	33''b	10. GY 7.0/ 5.0	LLG-14- 6°
Clear Green-Blue Gray	XLVIII	45''''d	10. B 6.0/ 1.5	UUC-12- 3°
Clear Payne's Gray	XLIX	49''''b	6.5 PB 5.2/ 1.5	U-11- 3°
Clear Windsor Blue	XXXV	49''	7.5 PB 4.6/ 6.0	U- 9- 7°
*Clear Yellow-Green	VI	31b	3.5 GY 8.0/ 7.5	L-17-11°
*Clove Brown	XL	17'''m	10. YR 2.8/ 1.5	SSO- 4- 2°
*Cobalt Green	XIX	37'b	3.5 G 7.0/ 5.0	GE-14-10°
Colonial Buff	XXX	21''d	5. Y 8.4/ 6.0	OY-18- 8°
*Columbia Blue	XXXIV	47''b	3.5 PB 5.5/ 8.0	UUC-12- 6°
Commelina Blue	XXI	51'	7.5 PB 5.0/15.0	U-10-12°
Congo Pink	XXVIII	7''b	1. YR 7.0/ 7.0	SSO-12- 9°
Coral Pink	XIII	5'd	10. R 7.0/ 7.5	SSO-14- 9°
Coral Red	XIII	5'	7.5 R 5.2/ 9.5	SSO-11-11°
Corinthian Pink	XXVII	3''d	5. R 7.2/ 4.5	RS-14- 6°
Corinthian Purple	XXXVIII	69''k	7.5 RP 3.4/ 6.0	R- 6- 4°
Corinthian Red	XXVII	3''	3.5 R 5.2/ 6.0	S-10- 6°
Cornflower Blue	XXI	53'	8.5 PB 4.8/15.0	U- 9-12°
Corydalis Green	XLI	29'''d	7.5 GY 8.0/ 3.0	L-15- 3°
Cossack Green	VI	33m	7.5 GY 4.4/ 3.0	GGL- 6- 3°
Cosse Green	V	29i	5. GY 5.8/ 9.0	LLY-12-11°
Cotinga Purple	XI	63K	7.5 P 3.5/ 8.0	VVM- 4- 9°
Courge Green	XVII	25'i	2.5 GY 5.8/ 4.5	YL-12- 5°
Court Gray	XLVII	29''''f	7.5 G 8.0/ 1.5	GE-14- 1°

Ridgway Color Name	Ridgway Notation		Hamly Notation		Villalobos Notation
Cream-Buff	XXX	19''d	2.5	Y 8.0/ 6.5	OOY-17- 6°
Cream Color	XVI	19'f	3.5	Y 8.5/ 5.5	OY-18- 8°
*Cress Green	XXXI	29''k	5.	GY 4.5/ 4.0	L- 7- 3°
Cyanine Blue	IX	51m	7.5	PB 3.0/ 6.0	U- 4-10°
Dahlia Carmine	XXVI	71'k	7.5	RP 3.2/ 6.0	MR- 6- 5°
Dahlia Purple	XII	67k	1.	P 2.8/ 7.0	MR- 4- 8°
Danube Green	XXXII	35''m	3.5	G 3.5/ 2.5	GGL- 5- 2°
Daphne Pink	XXXVIII	69''b	5.	RP 6.0/ 9.0	MR-12- 7°
Daphne Red	XXXVIII	69''	5.	RP 5.0/ 9.0	MR-10- 7°
Dark American Green	XLI	33'''k	2.5	G 3.5/ 2.5	G- 7- 3°
Dark Aniline Blue	X	55m	9.	PB 2.6/ 2.0	UUV- 2- 8°
Dark Anthracene Violet	XXV	61'm	5.	P 2.5/ 3.0	VVU- 4- 5°
Dark Bluish Glaucous	XLII	37'''b	5.	G 7.0/ 2.5	G-13- 3°
Dark Bluish Gray-Green	XLII	41'''k	7.5	BG 3.5/ 2.5	TC- 6- 3°
Dark Bluish Violet	X	57m	10.	PB 2.5/ 5.0	UUV-2/3-12°
Dark Cadet Blue	XXI	49'm	6.	PB 3.5/ 5.0	U- 5- 8°
Dark Chessylite Blue	XX	45'm	9.	B 3.0/ 5.0	CU- 5- 7°
Dark Cinnabar Green	XIX	39'k	10.	G 4.0/ 3.0	ET- 7- 4°
*Dark Citrine	IV	21m	5.	Y 4.2/ 4.0	OY- 6- 5°
Dark Corinthian Purple	XXXVIII	69''m	10.	RP 3.2/ 4.0	R- 4- 5°
Dark Cress Green	XXXI	29''m	5.	GY 3.5/ 3.5	L- 4- 2°
Dark Delft Blue	XLII	45'''m	2.5	PB 2.6/ 3.0	UUC- 4- 5°
Dark Diva Blue	XXI	51'k	7.5	PB 3.8/ 9.0	U- 6- 9°
Dark Dull Blue-Violet	XXXVI	53''k	10.	PB 3.2/ 4.0	VVU- 5- 6°
Dark Dull Bluish Violet (1)	XXIV	57*k	10.	PB 3.0/ 5.0	UV- 6- 6°
Dark Dull Bluish Violet (2)	XXXV	51''k	10.	PB 3.8/ 4.0	UUV- 6- 6°
Dark Dull Bluish Violet (3)	XXXVI	57''k	5.	P 3.0/ 5.0	VVU- 5- 6°
Dark Dull Violet-Blue	XXIV	53*k	9.	PB 3.6/ 4.0	UUV- 6- 6°
Dark Dull Violet-Blue	XXXV	55''k	10.	PB 3.2/ 4.0	UV- 5- 6°
Dark Dull Yellow-Green	XXXII	31''m	7.5	GY 3.5/ 3.0	L- 4- 2°
Dark Glaucous-Gray	XLVIII	37''''b	7.5	BG 5.0/ 1.5	T-11- 2°
Dark Gobelin Blue	XXXIV	43''k	2.5	B 3.8/ 3.0	C- 7- 3°
Dark Grayish Blue-Green	XLVIII	37''''k	7.5	BG 3.0/ 0.5	CCU- 4- 1°
Dark Grayish Blue-Violet	XXIV	55*k	10.	PB 3.4/ 4.0	UUV- 7- 7°
*Dark Grayish Brown	XLV	5''''k	2.5	R 3.8/ 1.5	S- 6- 2°
Dark Grayish Lavender	XLIII	57'''b	2.5	P 6.0/ 3.5	UV-11- 5°
Dark Grayish Olive	XLVI	21''''k	5.	Y 3.5/ 1.0	O- 6- 1°
Dark Green	XVIII	35'm	2.5	G 3.5/ 2.5	G- 7- 3°
Dark Green-Blue Gray	XLVIII	45''''	10.	B 4.2/ 3.5	UUC- 9- 3°
Dark Green-Blue Slate	XLVIII	45''''k	5.	PB 2.5/ 3.0	UUC- 3- 4°

190

Ridgway Color Name	Ridgway Notation		Hamly Notation		Villalobos Notation
Dark Greenish Glaucous	XLI	33'''b	10.	GY 6.5/ 4.0	GGL-13- 4°
Dark Greenish Olive	XXX	23''m	7.5	Y 3.8/ 3.0	YYO- 5- 2°
Dark Gull Gray	LIII	CG	N 4.3/		UUV-10- 2°
Dark Heliotrope Gray	L	65''''	5.	P 4.5/ 2.5	VVM- 9- 3°
Dark Heliotrope Slate	L	65''''k	5.	P 3.0/ 1.5	V- 4- 2°
Dark Hyssop Violet	XXXVI	59''k	7.5	P 3.5/ 4.0	V- 5- 6°
*Dark Indian Red	XXVII	3''m	5.	R 3.5/ 2.5	RS- 6- 3°
Dark Ivy Green	XLVII	25''''k	5.	GY 3.6/ 2.0	Y- 3- 1°
Dark Lavender	XLIV	61'''b	6.	P 6.5/ 6.0	V-11- 5°
*Dark Livid Brown	XXXIX	1'''k	2.5	R 3.6/ 2.0	R- 6- 3°
Dark Livid Purple	XXXVII	63''m	5.	RP 2.5/ 3.0	M- 4- 4°
Dark Madder Blue	XLIII	53'''k	8.5	PB 3.2/ 3.0	UUV- 5- 4°
Dark Madder Violet	XXV	63'm	5.	P 2.5/ 3.0	V- 4- 6°
Dark Maroon Purple	XXVI	71'm	2.5	RP 3.4/ 4.0	M- 5- 4°
Dark Medici Blue	XLVIII	41''''i	7.5	B 4.0/ 1.0	CU- 8- 3°
Dark Mineral Red	XXVII	1''m	2.5	R 3.5/ 2.0	MR- 5- 3°
Dark Mouse Gray	LI	15'''''k	7.5	YR 3.5/ 0.5	VM- 6- 1°
Dark Naphthalene Violet	XXXVII	61''m	10.	P 3.0/ 3.0	VM- 4- 5°
Dark Neutral Gray	LIII	NGk	N 3.1/		UUV- 4- 1°
Dark Nigrosin Violet	XXV	65'm	5.	P 2.2/ 3.0	VVM- 4- 5°
*Dark Olive	XL	21'''m	5.	Y 3.5/ 2.0	OOY- 3- 2°
Dark Olive-Buff	XL	21'''	5.	Y 6.2/ 4.5	OY-11- 5°
Dark Olive-Gray	LI	23'''''i	7.5	Y 4.0/ 1.5	OOS- 8- 1°
Dark Orient Blue	XXXIV	45''k	10.	B 4.0/ 3.0	UUC- 6- 6°
Dark Payne's Gray	XLIX	49''''k	6.5	PB 2.8/ 1.5	U- 4- 4°
Dark Perilla Purple	XXXVII	65''m	7.5	RP 3.0/ 3.5	MR- 3- 4°
Dark Plumbago Blue	XLIII	53'''b	10.	PB 5.6/ 3.0	UUV-12- 6°
Dark Plumbago Gray	L	61''''	7.5	P 4.5/ 1.5	UV- 9- 4°
Dark Plumbago Slate	L	61''''k	7.5	P 3.5/ 1.0	VVU- 4- 2°
Dark Plumbeous	LII	49'''''i	7.5	PB 4.2/ 1.0	UUV- 7- 2°
Dark Porcelain Green	XXXIII	39''k	2.5	G 3.8/ 2.0	GE- 7- 3°
Dark Purple-Drab	XLV	1''''i	8.	RP 3.6/ 2.0	R- 7- 3°
Dark Purplish Gray	LIII	67'''''k	10.	P 3.0/ 0.5	UUV- 4- 1°
Dark Quaker Drab	LI	1'''''k	5.	RP 3.5/ 1.0	VM- 6- 1°
Dark Russian Green	XLII	37'''k	7.5	G 3.5/ 2.0	E- 5- 4°
Dark Slate-Purple	XLIV	65'''k	2.5	RP 3.4/ 3.5	MMV- 6- 3°
Dark Slate-Violet (1)	XLIII	57'''k	1.	P 2.6/ 1.0	VVU- 6- 3°
Dark Slate-Violet (2)	XLIV	61'''k	7.5	P 3.8/ 2.5	V- 5- 4°
Dark Soft Blue-Violet	XXIII	55'k	9.	PB 3.8/ 6.0	UUV- 6- 8°
Dark Soft Bluish Violet	XXIII	57'k	10.	PB 3.0/ 8.5	UV- 6- 8°
Dark Sulphate Green	XIX	39'i	10.	G 4.6/ 4.0	ET- 8- 6°
Dark Terre Verte	XXXIII	41''k	2.5	BG 3.8/ 2.0	E- 6- 3°
*Dark Tyrian Blue	XXXIV	47''k	5.	PB 3.8/ 3.0	U- 6- 6°
Dark Varley's Gray	XLIX	57''''k	2.5	P 3.4/ 1.0	UV- 4- 4°
*Dark Vinaceous	XXVII	1''	2.5	R 4.6/ 5.0	RS-10- 5°
*Dark Vinaceous-Brown	XXXIX	5'''k	7.5	R 3.8/ 2.0	S- 6- 4°

191

Ridgway Color Name	Ridgway Notation		Hamly Notation		Villalobos Notation
*Dark Vinaceous-Drab	XLV	5''''i	2.5	R 3.8/ 1.5	S- 7- 3°
Dark Vinaceous-Gray	L	69''''	7.5	P 4.6/ 2.0	VVM- 9- 3°
Dark Vinaceous-Purple	XXXVIII	67''k	3.5	RP 3.4/ 5.0	R- 6- 4°
Dark Violet	X	59k	1.	P 3.2/ 7.0	UV- 4-10°
Dark Violet-Gray	LII	59'''''k	2.5	P 3.4/ 0.5	UV- 4- 2°
Dark Violet-Slate	XLIX	53''''k	8.5	PB 3.4/ 2.0	UUV-'4- 3°
Dark Viridian Green	VII	37k	6.	G 4.4/ 5.5	E- 6- 7°
*Dark Yellowish Green	XVIII	33'm	10.	GY 3.5/ 3.0	GGL- 6- 2°
Dark Yvette Violet	XXXVI	55''m	1.	P 2.8/ 2.0	VVU- 4- 4°
Dark Zinc Green	XIX	37'k	5.	G 4.5/ 3.0	GE- 7- 4°
Dauphin's Violet	XXIII	59'i	2.5	P 3.8/ 6.0	UV- 7- 9°
Dawn Gray	LII	35'''''d	2.5	B 6.8/ 0.5	TC-13- 1°
Deep Aniline Lilac	XXXV	53''b	10.	B 6.0/10.0	U-12- 8°
*Deep Blue-Violet	X	55i	9.	PB 2.8/14.0	UUV- 5-11°
Deep Bluish Glaucous	XLII	37'''d	7.5	BG 7.8/ 3.0	E-14- 4°
Deep Bluish Gray-Green	XLII	41'''i	5.	BG 4.2/ 2.5	T- 9- 3°
Deep Brownish Drab	XLV	9''''i	5.	YR 3.5/ 2.0	SSO- 8- 3°
Deep Brownish Vinaceous	XXXIX	5'''	5.	R 5.2/ 3.5	SSO- 9- 5°
Deep Cadet Blue	XXI	49'k	6.	PB 4.4/ 8.0	U- 6- 9°
*Deep Chicory Blue	XXIV	57*b	10.	PB 5.4/10.0	UUV-12-10°
*Deep Chrome	III	17b	9.	YR 6.8/12.0	OOY-17-12°
Deep Chrysolite Green	XXXI	27''	2.5	GY 6.0/ 4.5	LLY-13- 7°
Deep Colonial Buff	XXX	21''b	5.	Y 7.8/ 6.5	OY-16- 7°
Deep Corinthian Red	XXVII	3''i	6.5	R 4.4/ 4.0	S- 8- 5°
Deep Delft Blue	XLII	45'''k	2.5	PB 3.5/ 3.0	UUC- 5- 6°
Deep Dull Bluish Violet (1)	XXIV	57*i	10.	PB 3.8/ 9.0	UV- 8- 8°
Deep Dull Bluish Violet (2)	XXXV	51''i	9.	PB 4.5/ 4.5	UUV- 8- 7°
Deep Dull Bluish Violet (3)	XXXVI	57''i	2.5	P 3.8/ 6.0	UV- 7- 8°
Deep Dull Lavender	XLIV	61'''d	7.5	P 6.6/ 3.5	V-13- 6°
Deep Dull Violaceous Blue	XXII	51*k	7.5	PB 3.6/ 8.0	U- 6- 8°
Deep Dull Violet-Blue	XXXV	53''i	10.	PB 4.2/ 6.5	UV- 7- 7°
Deep Dull Yellow-Green (1)	XXXII	31''k	7.5	GY 4.5/ 4.0	L- 7- 4°
Deep Dull Yellow-Green (2)	XXXII	33''k	1.	G 4.4/ 3.0	LG- 7- 3°
Deep Dutch Blue	XLIII	49'''	6.5	PB 5.0/ 3.0	U-11- 5°
Deep Glaucous-Gray	XLVIII	37''''d	2.5	B 6.4/ 1.0	CCU-12- 2°
Deep Glaucous-Green	XXXIII	39''b	5.	G 6.8/ 4.0	GE-14- 6°
Deep Grape Green	XLI	25'''i	2.5	GY 5.0/ 3.0	YL-10- 3°
Deep Grayish Blue-Green	XLVIII	37''''i	7.5	BG 3.8/ 1.5	CCU- 8- 1°
Deep Grayish Lavender	XLIII	57''''d	2.5	P 6.2/ 3.0	VVU-13- 4°
Deep Grayish Olive	XLVI	21'''''i	5.	Y 4.0/ 1.0	O- 9- 1°

192

Ridgway Color Name	Ridgway Notation		Hamly Notation	Villalobos Notation
Deep Green-Blue Gray	XLVIII	45''''b	10. B 4.5/ 3.0	UUC-10- 4°
Deep Greenish Glaucous	XLI	33'''d	5. G 8.0/ 3.0	G-15- 4°
Deep Gull Gray	LIII	CGb	N 5.7/	UUV-12- 1°
Deep Heliotrope Gray	L	65''''b	5. P 5.2/ 2.5	V-10- 3°
Deep Hellebore Red	XXXVIII	71''i	7.5 RP 4.2/ 6.0	R- 8- 5°
Deep Hyssop Violet	XXXVI	59''i	5. P 3.8/ 7.0	VVU- 8- 6°
*Deep Lavender	XXXVI	59''d	2.5 P 6.5/ 6.0	UV-13- 8°
*Deep Lavender-Blue	XXI	53'b	8.5 PB 5.8/10.0	U-13-11°
Deep Lichen Green	XXXIII	37''d	1. G 7.5/ 3.0	GGL-15- 3°
Deep Livid Brown	XXXIX	1'''i	2.5 R 4.5/ 2.5	R- 8- 4°
Deep Livid Purple	XXXVII	63''k	5. RP 3.0/ 5.0	M- 6- 4°
Deep Madder Blue	XLIII	53'''i	8.5 PB 3.6/ 4.0	UUV- 7- 5°
*Deep Malachite Green	XXXII	35''	10. GY 5.6/ 5.0	G-12- 6°
Deep Medici Blue	XLVIII	41''''	7.5 B 4.8/ 1.0	CU-10- 2°
Deep Mouse Gray	LI	15''''i	7.5 YR 4.5/ 0.5	S- 8- 1°
Deep Neutral Gray	LIII	NGi	N 3.7/	UV- 7- 1°
*Deep Olive	XL	21'''k	5. Y 4.4/ 3.0	OY- 6- 3°
Deep Olive-Buff	XL	21'''b	5. Y 7.6/ 4.0	OOY-14- 4°
Deep Olive-Gray	LI	23'''''	7.5 Y 5.0/ 1.0	OOS-11- 1°
Deep Orient Blue	XXXIV	45''i	10. B 4.5/ 3.0	UUC- 9- 6°
Deep Payne's Gray	XLIX	49''''i	6.5 PB 3.8/ 2.0	U- 7- 3°
Deep Plumbago Blue	XLIII	53'''d	10. PB 6.4/ 2.5	UUV-13- 6°
Deep Plumbago Gray	L	61''''b	10. P 5.2/ 1.0	VVU-10- 2°
*Deep Plumbeous	LII	49'''''	7.5 PB 5.0/ 1.5	UUV-10- 2°
Deep Purplish Gray	LIII	67'''''i	10. P 3.8/ 1.0	UV- 7- 2°
Deep Purplish Vinaceous	XLIV	69'''	7.5 RP 4.8/ 5.0	R- 9- 4°
Deep Quaker Drab	LI	1'''''i	5. RP 4.0/ 1.0	M- 8- 1°
*Deep Rose-Pink	XII	71d	5. RP 6.6/12.0	MR-15-12°
*Deep Seafoam Green	XXXI	27''d	2.5 GY 8.0/ 5.0	YYL-17- 7°
Deep Slate-Blue	XLIII	49'''k	6.5 PB 3.4/ 2.0	U- 6- 4°
Deep Slate-Green	XLVII	33''''k	10. G 3.8/ 1.0	ET- 6- 1°
Deep Slate-Olive	XLVI	29''''k	3.5 G 4.0/ 1.0	LLY- 4- 1°
Deep Slate-Violet	XLIV	61'''i	7.5 P 4.2/ 3.0	V- 6- 5°
Deep Slaty Brown	L	69''''k	7.5 P 3.2/ 2.0	VVM- 5- 2°
Deep Soft Blue-Violet	XXIII	55'i	9. PB 4.0/ 8.0	UUV- 8- 8°
Deep Soft Bluish Violet	XXIII	57'i	10. PB 4.2/10.0	UV- 8- 8°
Deep Turtle Green	XXXII	31''	8.5 GY 6.0/ 7.0	LG-11- 6°
Deep Varley's Gray	XLIX	57''''i	2.5 P 4.0/ 1.5	UV- 7- 3°
Deep Vinaceous	XXVII	1''b	2.5 R 5.6/ 6.5	RS-12- 6°
Deep Vinaceous-Gray	L	69''''b	7.5 P 5.2/ 2.0	VVM-10- 3°
Deep Vinaceous-Lavender	XLIV	65'''d	2.5 RP 6.5/ 4.0	VVM-13- 5°
Deep Violet-Gray	LII	59'''''i	2.5 P 4.5/ 0.5	UV- 7- 2°
Deep Violet-Plumbeous	XLIX	53''''	8.5 PB 4.0/ 4.0	UV-10- 3°
*Deep Wedgwood Blue	XXI	51'd	7.5 PB 7.0/10.0	U-14-11°
Delft Blue	XLII	45'''i	2.5 PB 4.5/ 3.0	UUC- 8- 5°
Diamin-Azo Blue	XXXV	51''m	10. PB 3.0/ 3.0	UV- 4- 6°

193

Ridgway Color Name	Ridgway Notation		Hamly Notation		Villalobos Notation
*Diamine Brown	XIII	3′m	7.5	R 3.2/ 2.5	S- 5- 4°
Diamine Green	VII	37m	7.5	G 3.5/ 3.0	E- 5- 4°
Diva Blue	XXI	51′i	7.5	PB 4.5/11.0	U- 9-11°
Drab	XLVI	17′′′′	10.	YR 5.5/ 3.0	OOS-10- 3°
Drab-Gray	XLVI	17′′′′d	10.	YR 6.6/ 2.0	OOS-14- 2°
Dragons-blood Red	XIII	5′i	7.5	R 4.2/ 7.5	SSO- 9- 9°
*Dresden Brown	XV	17′k	10.	YR 3.8/ 4.0	O- 8- 5°
Duck Green	XIX	39′m	10.	G 3.2/ 2.0	T- 4- 5°
Dull Blackish Green	XLI	33′′′m	2.5	G 2.5/ 1.5	G- 5- 2°
Dull Blue-Green Black	XLVIII	41′′′′m	7.5	B 2.0/ 1.0	UUC- 3- 4°
Dull Blue-Violet (1)	XXIV	55*	9.	PB 4.5/12.0	UV- 8-10°
Dull Blue-Violet (2)	XXXVI	55′′i	1.	P 3.6/ 6.0	UV- 7- 7°
Dull Bluish Violet (1)	XXIV	57*	10.	PB 4.5/12.0	UV- 7-10°
Dull Bluish Violet (2)	XXXV	51′′	10.	B 4.5/ 6.0	UUV- 9- 7°
Dull Bluish Violet (3)	XXXVI	57′′	2.5	P 4.2/ 9.0	UV- 8- 8°
*Dull Citrine	XVI	21′k	5.	Y 4.5/ 4.5	Y- 8- 6°
Dull Dark Purple	XXVI	67′k	10.	P 3.3/ 9.0	MMV- 6- 5°
Dull Dusky Purple	XXVI	67′m	10.	P 2.6/ 3.0	VVM- 4- 4°
Dull Greenish Black (1)	XLVII	29′′′′m	3.5	G 3.5/ 0.5	LLG- 3- 1°
Dull Greenish Black (2)	XLVII	33′′′′m	10.	G 3.0/ 0.5	TC- 5- 1°
*Dull Green Yellow	XVII	27′	5.	GY 6.8/ 9.0	LLY-15- 9°
Dull Indian Purple	XLIV	69′′′i	5.	RP 3.8/ 3.5	M- 8- 4°
Dull Lavender	XLIV	61′′′′f	10.	P 7.4/ 3.0	VVM-15- 5°
*Dull Magenta Purple	XXVI	67′i	10.	P 4.5/10.5	M- 8- 6°
Dull Opaline Green	XIX	37′f	3.5	G 8.0/ 2.0	G-17- 7°
Dull Purplish Black	L	65′′′′m	5.	P 2.6/ 1.0	VVM- 4- 2°
Dull Violet-Black (1)	XLIV	61′′′m	7.5	P 3.0/ 1.0	V- 4- 3°
Dull Violet-Black (2)	XLIX	53′′′′m	8.5	PB 2.8/ 1.5	UUV- 3- 2°
Dull Violet-Black (3)	L	61′′′′m	7.5	P 2.5/ 0.5	VVU- 3- 2°
Dull Violaceous Blue	XXII	51*	7.5	PB 4.8/12.5	U- 9-11°
Dull Violet-Blue	XXIV	53*	8.5	PB 4.6/12.0	U- 8- 9°
Dull Violet-Blue	XXXV	53′′	10.	PB 4.2/ 8.0	UV- 8- 7°
Dusky Auricula Purple	XXVI	69′m	2.5	RP 2.2/ 3.0	MMV- 4- 4°
Dusky Blue	XXII	49*m	6.5	PB 4.4/ 4.0	U- 4- 8°
Dusky Blue-Green	XXXIII	39′′m	5.	BG 3.4/ 1.5	E- 6- 2°
Dusky Bluish Green	XXXIII	41′′m	2.5	BG 3.2/ 1.0	E- 5- 2°
Dusky Blue-Violet (1)	XXIII	57′m	10.	PB 2.8/ 4.0	UUV- 5- 6°
Dusky Blue-Violet (2)	XXIV	55*m	10.	PB 2.4/ 2.0	UUV- 3-10°
Dusky Brown	XLV	1′′′′k	8.	RP 3.2/ 1.5	R- 6- 2°
*Dusky Drab	XLV	9′′′′k	5.	YR 3.5/ 1.0	SSO- 6- 2°
Dusky Dull Bluish Green	XLII	41′′′m	7.5	BG 2.6/ 2.0	CCU- 3- 2°
Dusky Dull Green	XLII	37′′′m	7.5	G 2.5/ 1.0	ET- 4- 3°
Dusky Dull Violet (1)	XXXVI	57′′m	5.	P 3.0/ 4.0	VVU- 4- 5°
Dusky Dull Violet (2)	XXXVI	59′′m	7.5	P 3.5/ 5.0	V- 3- 4°
Dusky Dull Violet-Blue	XXXV	53′′m	10.	PB 3.2/ 3.0	UV- 4- 6°
Dusky Green	XXXIII	37′′m	5.	G 3.5/ 1.5	G- 6- 1°
Dusky Green-Blue (1)	XX	43′m	10.	BG 3.4/ 3.0	C- 5- 4°
Dusky Green-Blue (2)	XXXIV	43′′m	5.	B 3.0/ 2.5	C- 4- 3°

194

Ridgway Color Name	Ridgway Notation		Hamly Notation	Villalobos Notation
Dusky Green-Gray	LII	35′′′′′k	5. G 3.0/ 0.2	UUC- 4- 1°
Dusky Greenish Blue	XX	47′m	5. PB 3.0/ 7.0	UUC- 5- 8°
Dusky Neutral Gray	LIII	NGm	N 2.8/	UUV- 2- 1°
Dusky Olive-Green	XLI	25′′′m	2.5 GY 3.4/ 2.0	L- 4- 2°
Dusky Orient Blue	XXXIV	45′′m	10. B 3.5/ 3.0	CU- 5- 5°
Dusky Purplish Gray	LIII	67′′′′′m	10. P 2.5/ 0.5	UUV- 2- 2°
Dusky Slate-Blue	XLIII	49′′′m	7.5 PB 2.4/ 1.5	U- 4- 4°
Dusky Slate-Violet	XLIII	57′′′m	2.5 P 2.5/ 1.0	VVU- 4- 4°
Dusky Violet	XXIII	59′m	2.5 P 3.4/ 2.0	UV- 4- 6°
Dusky Violet-Blue (1)	XXIII	55′m	9. PB 3.0/ 3.5	UUV- 5- 6°
Dusky Violet-Blue (2)	XLIII	53′′′m	8.5 PB 2.4/ 1.5	UUV- 3- 4°
Dusky Yellowish Green	XLI	29′′′m	5. GY 3.4/ 2.5	GGL- 4- 2°
Dutch Blue	XLIII	49′′′b	8.5 PB 6.0/ 3.0	U-12- 5°
Ecru-Drab	XLVI	13′′′′d	6. YR 6.6/ 2.0	SO-13- 3°
Ecru-Olive	XXX	21′′i	5. Y 5.8/ 5.0	OY-11- 5°
Elm Green	XVII	27′m	5. GY 3.8/ 3.0	LLY- 6- 1°
*Emerald Green	VI	35	10. GY 7.5/ 9.0	GGL-14-12°
			or	
			2.0 G 6.7/10.0	
Empire Green	XXXII	33′′m	1. G 3.5/ 2.0	GGL- 5- 2°
Empire Yellow	IV	21b	4. Y 8.2/10.0	YYO-18-12°
Endive Blue	XLIII	49′′′d	8.5 PB 7.0/ 5.0	UUV-13- 4°
English Red	II	7i	10. R 4.2/10.0	SSO- 9-11°
Eosine Pink	I	1d	2.5 R 6.0/12.0	R-15-12°
Etain Blue	XX	43′f	10. BG 7.8/ 2.5	T-16- 4°
Ethyl Green	VII	41i	7.5 BG 4.0/ 7.0	T- 6-10°
Eton Blue	XXII	49*k	6.5 PB 3.5/ 8.0	UUC- 6- 9°
Etruscan Red	XXVII	5′′	7.5 R 5.0/ 4.5	SSO-10- 6°
Eugenia Red	XIII	1′	2.5 R 5.5/ 8.0	RS-10- 8°
Eupatorium Purple	XXXVIII	67′′	2.5 RP 5.5/ 8.0	M-11- 5°
Fawn Color	XL	13′′′	5. YR 4.6/ 3.0	OOS-10- 4°
Ferruginous	XIV	9′i	1. YR 4.5/ 8.0	SO-10- 7°
Flame Scarlet	II	9	10. R 6.0/14.0	SO-13-12°
*Flax-flower Blue	XXI	51′b	7.5 PB 6.0/11.0	U-13-11°
Flesh Color	XIV	7′d	2.5 R 7.0/ 6.5	SO-16- 9°
*Flesh Ocher	XIV	9′b	2.5 YR 6.6/ 8.0	SO-15- 9°
*Flesh Pink	XIII	5′f	10. R 8.0/ 5.5	S-16- 8°
*Fluorite Green	XXXII	33′′	1. G 6.4/ 6.0	GGL-11- 5°
Fluorite Violet	XI	61m	5. P 2.5/ 5.0	V- 3- 8°
*Forest Green	XVII	29′m	7.5 GY 4.5/ 3.5	L- 7- 4°
Forget-me-not Blue	XXII	51*b	7.5 PB 6.2/10.0	UUC-13-10°
French Gray	LII	49′′′′′f	7.5 PB 6.8/ 1.5	UV-15- 3°
French Green	XXXII	35′′i	3.5 G 4.6/ 3.5	G-10- 5°
Fuscous	XLVI	13′′′′k	5. YR 3.5/ 1.0	SO- 6- 2°
*Fuscous-Black	XLVI	13′′′′m	5. YR 2.6/ 0.5	S- 4- 2°
Garnet Brown	I	3k	6. R 3.0/10.0	RS- 4-12°
Gendarme Blue	XXII	47*k	5. PB 4.0/ 6.0	UUC- 6- 9°
Gentian Blue	XXI	53′i	8.5 PB 4.0/10.0	U- 7-11°
Geranium Pink	I	3d	4. R 6.4/11.0	RS-15-12°

Ridgway Color Name	Ridgway Notation		Hamly Notation		Villalobos Notation
Glass Green	XXXI	29"d	5.	GY 8.0/ 4.5	YYL-17- 7°
Glaucous	XLI	29'''f	7.5	GY 8.0/ 2.0	LG-16- 4°
Glaucous-Blue	XXXIV	43"b	2.5	B 6.4/ 4.0	C-12- 3°
*Glaucous-Gray	XLVIII	37''''f	2.5	B 7.5/ 1.0	CU-14- 1°
Glaucous-Green	XXXIII	39''d	5.	G 7.4/ 2.0	GE-16- 4°
Gnaphalium Green	XLVII	29'''d	5.	G 7.5/ 1.5	G-13- 1°
Gobelin Blue	XXXIV	43"i	5.	B 3.8/ 3.0	C- 9- 3°
Grape Green	XLI	25'''	2.5	GY 5.8/ 4.5	YYL-11- 4°
*Grass Green	VI	33k	7.5	GY 5.2/ 4.0	L- 8- 4°
Grayish Blue-Green	XLVIII	37''''	7.5	BG 4.5/ 1.5	TC-10- 2°
Grayish Blue-Violet (1)	XXIV	55*i	9.	PB 4.0/ 9.0	UUV- 9- 8°
Grayish Blue-Violet (2)	XXXV	51"b	7.5	PB 5.5/ 8.0	U-12- 8°
Grayish Lavender	XLIII	57'''f	5.	P 7.2/ 2.0	VVU-15- 3°
Grayish Olive	XLVI	21''''	5.	Y 4.8/ 2.5	O-10- 2°
Grayish Violaceous Blue	XXII	51*i	7.5	PB 4.6/ 8.0	U- 8- 9°
Grayish Violet-Blue	XXIV	53*i	10.	PB 4.5/ 6.0	UUV- 8- 8°
Green-Blue Slate	XLVIII	45''''i	10.	B 3.8/ 2.0	U- 7- 4°
*Green-Yellow	V	27b	1.	GY 8.2/ 9.0	LLY-18-12°
Greenish Glaucous	XLI	33'''f	10.	GY 8.2/ 2.0	GGL-16- 3°
Greenish Glaucous-Blue	XLII	41'''b	6.5	BG 6.0/ 3.0	T-12- 3°
Greenish Slate-Black	XLVIII	37''''m	7.5	BG 2.5/ 0.5	UUC- 3- 3°
*Greenish Yellow	V	25	10.	Y 7.8/10.0	YL-17-12°
				or	
			1.8	GY 8.2/11.0	
*Grenadine	II	7b	9.	R 6.0/12.0	SSO-15-12°
Grenadine Pink	II	7d	9.	R 7.0/10.0	SSO-16-12°
*Grenadine Red	II	7	8.5	R 5.4/13.0	SSO-12-12°
				or	
			10.	R 5.5/17.5	
Guinea Green	VII	39i	2.5	BG 4.2/ 8.0	ET- 7-10°
Gull Gray	LIII	CGd	N 6.1/		UUV-14- 1°
*Haematite Red	XXVII	5"m	7.5	R 3.4/ 2.0	RS- 6- 3°
Haematoxylin Violet	XXV	61'i	3.5	P 3.6/ 7.0	VVU- 7- 8°
*Hair Brown	XLVI	17''''i	10.	YR 4.4/ 1.0	OOS- 9- 2°
Hathi Gray	LII	35'''''b	5.	G 6.5/ 0.5	TC-11- 1°
*Hay's Blue	IX	53k	7.5	PB 3.2/10.5	U- 5-11°
*Hay's Brown	XXXIX	9'''k	10.	R 4.5/ 2.0	SSO- 5- 3°
Hay's Green	XVIII	33'k	10.	GY 4.5/ 4.0	GGL- 8- 4°
Hay's Lilac	XXXVII	63''d	5.	P 6.4/ 7.0	V-12- 7°
Hay's Maroon	XIII	1'm	2.5	R 2.5/ 2.0	RS- 5- 4°
*Hay's Russet	XIV	7'k	10.	R 4.0/ 5.0	SSO- 7- 6°
Hazel	XIV	11'k	2.5	YR 4.4/ 5.5	OOS- 8- 8°
Heliotrope Gray	L	65''''d	5.	P 5.5/ 2.0	V-12- 3°
Heliotrope Slate	L	65''''i	5.	P 3.8/ 2.0	V- 6- 3°
Hellebore Green	XVII	25'm	2.5	GY 4.5/ 3.0	YL- 6- 2°
Hellebore Red	XXXVIII	71''	7.5	RP 4.8/ 9.0	R-11- 7°
*Helvetia Blue	IX	51k	7.5	PB 3.2/10.0	U- 5-10°

Ridgway Color Name	Ridgway Notation		Hamly Notation		Villalobos Notation
Hermosa Pink	I	1f	2.5	R 7.8/ 8.0	R-17-12°
*Hessian Brown	XIII	5'm	10.	R 2.5/ 2.5	S- 5- 5°
Honey Yellow	XXX	19''	2.5	Y 6.5/ 7.0	OOY-14- 6°
Hortense Blue	XXII	47*m	5.	PB 3.5/ 5.0	UUC- 4- 8°
*Hortense Violet	XI	61b	5.	P 6.0/12.0	VVU-12-12°
*Hyacinth Blue	X	55k	9.	PB 2.6/ 8.0	UUV- 4-10°
Hyacinth Violet	XI	61i	5.	P 3.0/10.0	V- 4-12°
Hydrangea Pink	XXVII	5''f	7.5	R 8.0/ 5.0	S-16- 5°
Hydrangea Red	XXVII	1''i	2.5	R 4.2/ 5.0	RS- 8- 5°
Hyssop Violet	XXXVI	59''	5.	P 5.0/ 5.5	VVU- 9- 6°
Indian Lake	XXVI	71'i	7.5	RP 3.8/ 7.0	R- 8- 6°
Indian Purple	XXXVIII	67''m	7.5	RP 3.0/ 3.0	R- 5- 4°
*Indian Red	XXVII	3''k	5.	R 4.0/ 4.0	RS- 7- 4°
Indigo Blue	XXXIV	47''m	5.	PB 3.0/ 3.0	U- 5- 6°
Indulin Blue	XXII	51*m	7.5	PB 3.0/ 3.5	UUV- 4- 7°
Invisible Green	XIX	41'm	7.5	BG 3.0/ 2.5	TC- 5- 4°
Iron Gray	LI	23'''''k	7.5	Y 3.8/ 1.0	N-5/
*Isabella Color	XXX	19''i	2.5	Y 5.8/ 4.5	OOY-11- 4°
Italian Blue	VIII	43	5.	B 5.0/10.0	C-10-11°
				or	
			3.5	B 5.2/11.0	
Ivory Yellow	XXX	21''f	5.	Y 8.5/ 3.0	OOY-19- 6°
Ivy Green	XXXI	25''m	10.	Y 3.5/ 3.0	YYL- 4- 2°
Jade Green	XXXI	27''k	2.5	GY 4.2/ 3.0	YL- 8- 3°
Japan Rose	XXVIII	9''b	5.	YR 6.6/ 4.5	SO-14- 7°
Jasper Green	XXXIII	37''i	7.5	G 5.0/ 3.0	E- 9- 4°
Jasper Pink	XIII	3'd	6.	R 6.5/ 8.5	S-14- 9°
Jasper Red	XIII	3'	5.	R 4.8/10.5	S-10-10°
Javel Green	V	27i	1.	GY 6.8/ 6.0	YL-14-12°
Jay Blue	XXII	47*i	5.	PB 4.5/ 7.0	UUC- 8- 8°
Jovence Blue	XX	43'k	10.	BG 4.0/ 4.0	TC-5/6-7°
*Kaiser Brown	XIV	9'k	2.5	YR 3.8/ 6.0	SO- 8- 5°
Kildare Green	XXXI	29''b	5.	GY 7.2/ 5.0	LLY-14- 5°
Killarney Green	XVIII	35'i	2.5	G 5.0/ 5.0	G-10- 5°
King's Blue	XXII	47*b	2.5	PB 5.8/ 8.0	CU-12-10°
Kronberg's Green	XXXI	25''k	10.	Y 4.5/ 4.0	YYL- 7- 3°
Laelia Pink	XXXVIII	67''d	2.5	RP 6.4/ 9.0	MR-13- 6°
La France Pink	I	3f	2.5	R 7.5/12.0	RS-17-11°
Lavender	XXXVI	59''f	2.5	P 7.6/ 5.0	VVU-16- 5°
*Lavender-Blue	XXI	53'd	8.5	PB 6.5/ 9.0	U-14-11°
Lavender-Gray	XLIII	49'''f	9.	PB 7.5/ 4.0	UUV-15- 4°
*Lavender-Violet	XXV	61'b	2.5	P 6.0/10.0	UV-13-11°
Leaf Green	XLI	29'''k	5.	GY 4.0/ 2.5	LLG- 7- 3°
Leitch's Blue	VIII	47i	2.5	B 4.0/10.0	CU- 6-12°
Lemon Chrome	IV	21	4.	Y 7.8/10.0	YYO-17-12°
				or	
			5.2	Y 8.0/11.0	
*Lemon Yellow	IV	23	7.5	Y 8.0/10.5	Y-18-12°
*Lettuce Green	V	29k	4.	GY 5.8/ 5.5	LLY-10- 6°

Ridgway Color Name	Ridgway Notation		Hamly Notation		Villalobos Notation
Lichen Green	XXXIII	37″f	3.5	G 8.2/ 2.0	GE-17- 4°
Light Alice Blue	XXXIV	45″d	2.5	PB 7.0/ 5.0	UUC-14- 6°
Light Amparo Blue	IX	51d	6.	PB 7.0/10.0	UUC-13-12°
Light Amparo Purple	XI	63d	5.	P 7.0/10.0	V-14-12°
Light Bice Green	XVII	29′i	7.5	GY 5.6/ 4.5	L-10- 5°
Light Blue-Green	VII	39d	7.5	G 7.0/ 7.0	E-15-12°
Light Blue-Violet	X	55b	8.5	PB 5.4/14.0	U/UUV-12-12°
Light Bluish Violet	X	57b	9.	PB 5.5/11.0	UUV-11-12°
*Light Brownish Drab	XLV	9″″b	6.	YR 5.4/ 2.0	SO-11- 3°
*Light Brownish Olive	XXX	19″k	2.5	Y 4.6/ 3.5	O- 9- 2°
Light Brownish Vinaceous	XXXIX	5‴d	2.5	R 7.0/ 2.0	RS-15- 4°
*Light Buff	XV	17′f	1.	Y 8.5/ 4.5	O-19- 6°
Light Cadet Blue	XXI	49′b	6.	PB 6.0/11.0	UUC-13-11°
Light Cadmium	IV	19	2.5	Y 7.5/12.0	OOY-18-11°
Light Campanula Blue	XXIV	55*d	8.5	PB 6.6/10.0	U-15-10°
Light Celandine Green	XLVII	33″″d	5.	BG 6.5/ 1.0	C-13- 1°
Light Cendre Green	VI	35d	10.	GY 8.0/ 7.0	GGL-15-10°
Light Cerulean Blue	VIII	45b	10.	B 6.0/ 8.0	CCU-13-12°
*Light Chalcedony Yellow	XVII	25′d	10.	Y 8.4/ 6.0	YYL-18- 8°
Light Chicory Blue	XXIV	57*f	10.	PB 7.4/ 6.0	UUV-16- 9°
Light Cinnamon-Drab	XLVI	13″″b	5.	YR 5.8/ 2.5	SO-12- 3°
Light Columbia Blue	XXXIV	47″d	3.5	PB 6.5/ 6.0	UUC-13- 6°
Light Congo Pink	XXVIII	7″d	2.5	YR 8.0/ 5.0	SSO-15- 8°
Light Coral Red	XIII	5′b	7.5	R 6.0/10.0	SSO-13-15°
Light Corinthian Red	XXVII	3″b	5.	R 6.2/ 6.5	S-12- 7°
Light Cress Green	XXXI	29″i	5.	GY 5.2/ 4.0	LLY-10- 3°
Light Danube Green	XXXII	35″k	5.	G 4.0/ 3.0	GE- 7- 4°
*Light Drab	XLVI	17″″b	10.	YR 5.8/ 2.0	OOS-12- 3°
*Light Dull Bluish Violet	XXXVI	57″b	1.	P 5.6/ 8.0	UV-12- 6°
Light Dull Glaucous-Blue	XLII	41‴d	2.5	B 6.8/ 3.0	TC-14- 4°
Light Dull Green-Yellow	XVII	27′d	3.5	GY 7.6/ 4.5	YL-17- 8°
Light Elm Green	XVII	27′k	6.	GY 4.8/ 4.0	LLY- 8- 2°
Light Fluorite Green	XXXII	33″d	7.5	GY 8.0/ 4.0	L-16- 5°
Light Forget-me-not Blue	XXII	51*d	5.	PB 6.8/ 8.0	UUC-15- 9°
Light Glaucous-Blue	XXXIV	43″d	5.	B 7.6/ 4.0	C-14- 3°
Light Grape Green	XLI	25‴b	2.5	GY 7.0/ 4.5	YL-14- 3°
Light Grayish Blue-Violet	XXXV	51″d	6.5	PB 7.2/ 7.0	U-14- 8°
*Light Grayish Olive	XLVI	21″″b	5.	Y 5.8/ 2.0	O-12- 2°
Light Grayish Vinaceous	XXXIX	9‴d	10.	R 6.8/ 2.0	S-14- 5°
*Light Grayish Violet-Blue	XXIV	53*b	8.5	PB 5.8/10.0	U-13-10°
*Light Greenish Yellow	V	25b	10.	Y 8.0/ 9.0	YYL-18-11°

198

Ridgway Color Name	Ridgway Notation		Hamly Notation	Villalobos Notation
*Light Green-Yellow	V	27d	1. GY 8.4/ 7.0	YL-18-10°
Light Gull Gray	LIII	CGf	N 6.5/	UUV-15- 1°
Light Heliotrope-Gray	L	65''''f	2.5 RP 6.4/ 2.0	VVU-14- 3°
Light Hellebore Green	XVII	25'k	2.5 GY 5.0/ 3.0	YL- 9- 2°
*Light Hortense Violet	XI	61d	5. P 6.0/10.0	V-13-12°
Light Hyssop Violet	XXXVI	59''b	2.5 P 5.6/ 7.0	UV-11- 6°
Light Jasper Red	XIII	3'b	5. R 5.5/10.5	S-12-10°
Light King's Blue	XXII	47*d	1. PB 7.5/ 4.0	CU-15- 8°
Light Lavender-Blue	XXI	53'f	8.5 PB 7.4/ 8.0	U-16-12°
*Light Lavender-Violet	XXV	61'd	1.5 P 6.0/ 7.0	UV-14-11°
Light Lobelia Violet	XXXVII	61''d	6. P 6.5/ 7.0	V-13- 7°
Light Lumiere Green	XVII	29'd	7.5 GY 8.0/ 4.5	L-17- 7°
Light Mallow Purple	XII	67d	1. P 6.0/12.0	VM-14- 2°
*Light Mauve	XXV	63'd	2.5 P 7.0/ 8.0	VVU-15-10°
Light Medici Blue	XLVIII	41''''d	10. B 5.4/ 2.0	UUC-12- 3°
Light Methyl Blue	VIII	47b	2.5 PB 5.5/10.0	CU-11-12°
Light Mineral Gray	XLVII	25''''f	5. GY 8.4/ 1.0	SSO-14- 1°
Light Mouse Gray	LI	15'''''b	7.5 YR 6.0/ 1.0	VM-11- 1°
Light Neropalin Blue	XXII	49*d	3.5 PB 7.0/ 4.0	UUC-15- 9°
Light Neutral Gray	LIII	NGb	N 5.2/	UV-12- 1°
Light Niagara Green	XXXIII	41''d	5. BG 7.0/ 5.0	ET-15- 4°
Light Ochraceous-Buff	XV	15'd	8.5 YR 7.2/ 8.0	O-17- 7°
Light Ochraceous-Salmon	XV	13'd	8. YR 8.0/ 5.5	OOS-16- 7°
Light Olive-Gray	LI	23'''''d	7.5 Y 7.0/ 1.0	O-14- 1°
*Light Orange-Yellow	III	17d	1. Y 7.5/ 9.0	OOY-17-10°
*Light Oriental Green	XVIII	33'b	2.5 G 6.6/ 6.0	GGL-14- 7°
Light Paris Green	XVIII	35'd	1. G 7.5/ 4.0	GGL-16- 7°
Light Payne's Gray	XLIX	49''''d	6.5 PB 5.8/ 1.5	U-12- 3°
Light Perilla Purple	XXXVII	65''i	2.5 RP 4.0/ 5.0	M- 8- 5°
Light Phlox Purple	XI	65d	10. P 6.0/ 9.5	VVM-13-12°
Light Pinkish Cinnamon	XXIX	15''d	1. YR 8.0/ 5.0	OOS-16- 7°
Light Pinkish Lilac	XXXVII	65''f	2.5 RP 7.8/ 5.5	M-16- 4°
Light Plumbago Gray	L	61''''f	10. P 6.4/ 1.0	VVU-14- 3°
Light Porcelain Green	XXXIII	39''	5. BG 5.5/ 4.0	ET-10- 4°
Light Purple-Drab	XLV	1''''b	7.5 RP 5.4/ 2.0	R-11- 3°
Light Purplish Gray	LIII	67'''''b	10. P 4.8/ 0.5	VVM-10- 1°
Light Purplish Vinaceous	XXXIX	1'''d	10. RP 7.0/ 3.5	R-14- 6°
Light Quaker Drab	LI	1'''''b	5. RP 5.4/ 1.0	VVM-10- 2°
Light Rosolane Purple	XXVI	69'b	3.5 RP 5.8/13.0	M-14- 9°
Light Russet-Vinaceous	XXXIX	9'''b	1. YR 6.4/ 4.0	SSO-12- 6°
Light Salmon-Orange	II	11d	5. YR 7.0/ 7.0	SO-17-10°
*Light Seal Brown	XXXIX	9'''m	10. R 2.6/ 1.0	S- 4- 3°
Light Sky Blue	XX	47'f	2.5 PB 7.4/ 4.0	UUC-16-12°
*Light Soft Blue-Violet	XXIII	55'b	9. PB 6.0/10.0	UUV-13-12°
*Light Squill Blue	XX	45'd	2.5 PB 6.8/ 6.0	CU-14-11°

Ridgway Color Name	Ridgway Notation		Hamly Notation		Villalobos Notation
Light Sulphate Green	XIX	39′b	2.5	BG 6.2/ 6.0	ET-13-10°
Light Terre Verte	XXXIII	41″	6.	BG 5.2/ 3.4	T- 9- 5°
Light Turtle Green	XXXII	31″d	4.	GY 8.0/ 4.5	LLY-15- 5°
Light Tyrian Blue	XXXIV	47″	5.	PB 4.5/ 3.0	U-10- 5°
Light Varley's Gray	XLIX	57″″b	2.5	P 5.8/ 2.0	UV-10- 2°
Light Vinaceous-Cinnamon	XXIX	13″d	7.5	YR 8.0/ 5.0	OOS-15- 7°
Light Vinaceous-Drab	XLV	5″″b	2.5	R 5.4/ 2.0	S-11- 2°
Light Vinaceous-Fawn	XL	13″″d	5.	YR 7.4/ 2.5	SSO-15- 5°
Light Vinaceous-Gray	L	69″″f	2.5	RP 6.5/ 1.5	VVM-14- 3°
Light Vinaceous-Lilac	XLIV	69″′d	2.5	RP 6.5/ 5.0	M-14- 4°
Light Vinaceous-Purple	XLIV	65″′b	2.5	RP 5.2/ 4.5	VM-10- 5°
Light Violet	X	59b	9.	PB 6.0/12.0	UUV-11-12°
Light Violet-Blue	IX	53b	8.5	PB 5.0/12.0	U-11-12°
Light Violet-Gray	LII	59″″″b	2.5	P 5.6/ 0.5	UV-13- 2°
Light Violet-Plumbeous	XLIX	53″″d	8.5	PB 5.5/ 3.0	UUV-12- 2°
Light Viridine Green	VI	33f	3.5	GY 8.5/ 5.0	L-19-10°
*Light Viridine Yellow	V	29d	1.5	GY 8.2/ 7.0	LLY-18-12°
Light Windsor Blue	XXXV	49″b	7.5	PB 5.6/ 8.0	U-12- 8°
Light Wistaria Blue	XXIII	57′d	9.	PB 7.0/ 8.0	UUV-14-12°
Light Wistaria Violet	XXIII	59′d	10.	PB 7.0/ 8.0	UUV-14-12°
Light Yellow-Green	VI	31d	5.	GY 8.0/ 7.0	LLY-17-10°
*Light Yellowish Olive	XXX	23″i	7.5	Y 5.8/ 6.0	YYO-10- 5°
Lilac	XXV	65′d	5.	P 7.0/ 5.0	V-15-10°
Lilac-Gray	LII	59″″″f	2.5	P 7.0/ 1.0	UV-15- 2°
Lily Green	XLVII	33″″i	2.5	BG 5.0/ 1.0	TC- 8- 3°
Lime Green	XXXI	25″	10.	Y 6.5/ 5.0	YYL-14- 6°
Lincoln Green	XLI	25″′k	2.5	GY 4.0/ 2.0	YL- 6- 2°
Liseran Purple	XXVI	67′b	10.	P 5.5/ 9.0	VM-13-10°
Litho Purple	XXV	63′i	5.	P 3.5/ 9.0	U- 7- 8°
*Liver Brown	XIV	7′m	10.	R 3.0/ 3.5	SSO- 5- 4°
Livid Brown	XXXIX	1″′	3.	R 5.0/ 3.0	R-10- 4°
Livid Pink	XXVII	3″f	5.	R 8.0/ 3.0	R-16- 6°
Livid Purple	XXXVII	63″i	10.	P 3.6/ 6.0	VM- 7- 6°
Livid Violet	XXXVII	61″i	7.5	P 3.8/ 5.0	V- 6- 7°
Lobelia Violet	XXXVII	61″b	6.5	P 5.8/ 7.0	V-16- 6°
Lumiere Blue	XX	43′d	10.	BG 6.5/ 4.0	TC-15- 7°
*Lumiere Green	XVII	29′b	7.5	GY 7.5/ 6.0	L-16- 8°
Lyons Blue	IX	51i	6.	PB 3.4/12.0	UUC- 5-13°
Madder Blue	XLIII	53″′	10.	PB 5.0/ 3.0	UUV-10- 5°
Madder Brown	XIII	3′k	7.5	R 3.6/ 4.0	S- 6- 6°
Madder Violet	XXV	63′k	5.	P 2.8/ 4.5	VVU- 5- 7°
Magenta	XXVI	67′	10.	P 5.0/12.0	M-10- 7°
Mahogany Red	II	7k	10.	R 3.5/ 6.0	SSO- 6- 9°
*Maize Yellow	IV	19f	2.5	Y 8.5/ 6.0	OOY-19-12°
*Malachite Green	XXXII	35″b	10.	GY 6.8/ 5.0	LG-13- 5°
Mallow Pink	XII	67f	1.	P 7.0/10.0	VM-15-11°
Mallow Purple	XII	67b	1.	P 5.5/14.0	MMV-12-12°
Manganese Violet	XXV	63′	5.	P 4.5/11.0	V- 8- 9°

200

Ridgway Color Name	Ridgway Notation		Hamly Notation		Villalobos Notation
Marguerite Yellow	XXX	23"f	6.5	Y 8.4/ 3.0	YYO-19- 3°
Marine Blue	VIII	45m	10.	B 3.2/ 6.0	CCU- 4-12°
Maroon	I	3m	7.5	R 2.4/ 4.0	RS- 6- 3°
*Mars Brown	XV	13'm	5.	YR 3.4/ 3.5	SO- 6- 3°
Mars Orange	II	9i	10.	R 4.5/10.0	SO- 9-12°
Mars Violet	XXXVIII	71"m	10.	RP 3.0/ 2.5	R- 5- 3°
Mars Yellow	III	15i	7.5	YR 5.6/10.0	O-12-12°
Martius Yellow	IV	23f	7.5	Y 8.5/ 7.0	Y-19-12°
Massicot Yellow	XVI	21'f	6.	Y 8.4/ 5.0	Y-18- 6°
Mathews' Blue	XX	45'	2.5	PB 4.8/10.0	CU-10-11°
Mathews' Purple	XXV	65'	5.	P 4.0/ 9.0	VVM- 9- 8°
Mauve	XXV	63'b	2.5	P 6.0/10.0	VVU-12-10°
Mauvette	XXV	65'f	10.	P 8.0/ 6.0	V-16- 9°
Mazarine Blue	IX	49d	5.	PB 6.0/ 9.0	UUC-14-13°
Meadow Green	VI	35k	2.5	G 4.6/ 4.0	G- 8- 5°
Medal Bronze	IV	19m	2.5	Y 3.8/ 4.0	OOY- 6- 8°
Medici Blue	XLVIII	41''''b	7.	B 5.0/ 1.0	CCU-11- 2°
*Methyl Blue	VIII	47	2.5	PB 5.0/14.0	CU- 8-13°
Methyl Green	XIX	41'	5.	BG 5.4/ 7.0	T- 9- 9°
Microcline Green	XIX	39'f	2.5	BG 8.0/ 4.0	E-16- 7°
Mignonette Green	XXXI	25"i	10.	Y 3.5/ 5.0	YYL-12- 5°
*Mikado Brown	XXIX	13"i	5.	YR 5.0/ 5.0	OOS- 9- 6°
*Mikado Orange	III	13b	5.	YR 7.0/12.0	OOS-15-11°
Mineral Gray	XLVII	25''''d	5.	GY 7.4/ 2.0	OY-15- 1°
*Mineral Green	XVIII	31'	6.5	GY 6.5/ 8.0	L-14-11°
*Mineral Red	XXVII	1"k	2.5	R 3.6/ 4.0	R- 7- 4°
Montpellier Green	XXXIII	37"	7.5	G 5.5/ 4.0	E-11- 5°
Morocco Red	I	5k	7.5	R 3.4/ 6.0	S- 5- 8°
Motmot Blue	XX	43'	2.5	B 5.2/ 6.5	TC-10-10°
*Motmot Green	XVIII	35'	2.5	G 5.8/ 7.0	G-11- 8°
Mouse Gray	LI	15'''''	7.5	YR 5.0/ 1.0	SO- 9- 1°
*Mummy Brown	XV	17'm	10.	YR 3.4/ 3.0	SO- 4-12°
Mulberry Purple	XI	61k	5.	P 3.0/ 7.0	V- 4- 9°
Mustard Yellow	XVI	19'b	3.5	Y 7.8/ 9.0	OY-17-10°
Mytho Green	XLI	29'''b	7.5	GY 7.0/ 3.0	L-13- 3°
Myrtle Green	VII	41m	9.	BG 3.2/ 4.0	TC- 5- 6°
Naphthalene Violet	XXXVII	61"k	10.	P 3.2/ 5.0	VM- 5- 5°
Naphthalene Yellow	XVI	23'f	7.5	Y 8.5/ 5.0	Y-18- 8°
*Naples Yellow	XVI	19'd	3.5	Y 8.2/ 7.0	OY-17- 8°
Natal Brown	XL	13'''k	5.	YR 3.8/ 2.0	SO- 7- 3°
Navy Blue	XXI	53'm	8.6	PB 2.8/ 5.0	UUV- 5- 8°
Neropalin Blue	XXII	49*b	3.5	PB 6.0/ 8.5	UUC-13- 9°
Neutral Gray	LIII	NG	N 4.7/		UV-10- 1°
Neutral Red	XXXVIII	71"k	7.5	RP 3.5/ 5.5	R- 6- 5°
Neuvider Green	VII	37d	10.	GY 8.2/ 6.0	GGL-16-10°
Neva Green	V	29	5.	GY 7.6/10.0	LLG-16-12°
Niagara Green	XXXIII	41"b	2.5	BG 6.2/ 2.5	ET-14- 4°
Nickel Green	XXXIII	37"k	5.	G 4.0/ 2.5	G- 7- 2°
Night Green	VI	33	6.5	GY 7.2/12.0	LG-15-12°

201

Ridgway Color Name	Ridgway Notation		Hamly Notation		Villalobos Notation
Nigrosin Blue	XXXV	49″m	7.5	PB 4.2/ 3.5	UUV- 4- 5°
Nigrosin Violet	XXV	65′k	5.	P 2.8/ 6.0	VVM-5/6-6°
*Nile Blue	XIX	41′d	5.	BG 7.0/ 6.0	T-14- 9°
*Nopal Red	I	3i	4.	R 4.0/12.0	S- 8-11°
*Ochraceous-Buff	XV	15′b	1.	YR 8.0/ 6.0	O-16- 9°
*Ochraceous-Orange	XV	15′	8.5	YR 6.5/12.0	O-14- 8°
Ochraceous-Salmon	XV	13′b	5.	YR 7.4/ 7.0	OOS-15- 9°
*Ochraceous-Tawny	XV	15′i	8.5	YR 5.5/ 7.0	O-12- 8°
Ocher Red	XXVII	5″i	7.5	R 4.4/ 4.0	SSO- 8- 5°
*Oil Green	V	27k	4.	GY 5.0/ 4.5	LLY- 9- 5°
Oil Yellow	V	25i	10.	Y 6.0/ 6.0	YYL-13-11°
Old Gold	XVI	19′i	3.5	Y 5.5/ 6.0	OOY-13-10°
Old Rose	XIII	1′b	2.5	R 5.2/ 7.5	RS-12- 9°
Olivaceous Black (1)	XLVI	21‴′m	5.	Y 3.0/ 1.0	O- 4- 1°
Olivaceous Black (2)	XLVII	25‴′m	5.	GY 2.5/ 1.5	OOS- 4- 1°
Olivaceous Black (3)	LI	23‴‴m	7.5	Y 3.0/ 0.5	N-4/
Olive	XXX	21″m	5.	Y 3.8/ 3.0	OOY- 5- 2°
Olive-Brown	XL	17‴k	10.	YR 3.8/ 2.0	O- 6- 3°
Olive-Buff	XL	21‴d	5.	Y 8.0/ 3.5	OOY-16- 3°
*Olive-Citrine	XVI	21′m	5.	Y 3.8/ 3.5	Y- 5- 5°
Olive-Gray	LI	23‴‴b	7.5	Y 5.8/ 2.0	O-12- 1°
Olive-Green	IV	23m	7.5	Y 4.0/ 3.5	Y- 6- 7°
Olive Lake	XVI	21′i	5.	Y 5.5/ 4.5	YYO-11- 8°
Olive-Ocher	XXX	21″	5.	Y 6.5/ 6.5	OY-14- 9°
Olive-Yellow	XXX	23″	7.5	Y 6.8/ 7.0	Y-13- 8°
Olivine	XXXII	35″d	10.	G 7.5/ 4.0	LLG-16- 6°
Olympic Blue	XX	47′	3.5	PB 5.0/11.0	UUC-11-11°
Onion-skin Pink	XXVIII	11″b	5.	YR 7.0/ 6.0	SO-14- 8°
Ontario Violet	XXXVI	55″b	10.	PB 5.2/ 8.0	UUV-12- 8°
Opaline Green	VII	37f	6.5	GY 8.4/ 3.0	GGL-18-10°
Orange	III	15	5.	YR 6.6/12.0	O-16/17-12°
				or	
			5.2	R 6.7/14.0	
Orange-Buff	III	15d	8.5	YR 7.2/ 9.0	O-17-10°
Orange-Chrome	II	11	2.5	YR 6.0/15.0	SO-13-12°
*Orange-Cinnamon	XXIX	13″	5.	YR 5.8/ 7.0	OOS-11- 7°
Orange-Citrine	IV	19k	2.5	Y 4.8/ 6.0	OOY- 9- 9°
Orange-Pink	II	11f	2.5	YR 8.0/10.0	SO-18-10°
*Orange-Rufous	II	11i	2.5	R 5.0/10.0	SO-10-12°
Orange-Vinaceous	XXVII	5″b	5.	R 6.5/ 9.5	SSO-12- 7°
*Oriental Green	XVIII	33′	2.5	G 5.6/ 6.0	GGL-12- 8°
Orient Blue	XXXIV	45″	10.	B 5.0/ 5.0	UUC-10- 7°
Orient Pink	II	9f	10.	R 8.0/ 8.0	SSO-18-12°
Oural Green	XVIII	35′f	8.5	GY 8.5/ 3.0	GGL-17- 5°
Ox-blood Red	I	1k	5.	R 2.4/10.0	RS- 4- 8°
Oxide Blue	VIII	45i	10.	B 3.5/ 9.0	CCU- 6-12°
Pale Amaranth Pink	XII	69f	2.5	RP 7.0/10.0	MMV-16-12°
Pale Amparo Blue	IX	51f	4.	PB 7.5/ 4.0	UUC-16-12°
Pale Amparo Purple	XI	63f	7.5	P 7.0/ 8.0	V-16-12°

Ridgway Color Name	Ridgway Notation		Hamly Notation		Villalobos Notation
Pale Aniline Lilac	XXXV	53″f	10.	B 7.5/ 6.0	UV-15- 8°
Pale Blue (Ethyl Blue)	VIII	45f	3.5	B 8.0/ 3.0	C-17-11°
Pale Blue-Green	VII	39f	2.5	G 8.4/ 2.0	GE-17-11°
Pale Blue-Violet	X	55d	8.5	PB 6.4/10.0	U-14-12°
Pale Bluish Lavender	XXXVI	57″f	2.5	P 8.4/ 3.0	UV-15- 7°
Pale Bluish Violet	X	57d	8.5	PB 7.0/10.0	UUV-14-12°
Pale Brownish Drab	XLV	9″″d	6.	YR 7.0/ 1.0	S-14- 3°
Pale Brownish Vinaceous	XXXIX	5‴f	2.5	R 8.0/ 2.5	RS-16- 4°
*Pale Cadet Blue	XXI	49′d	5.	PB 6.5/ 5.0	UUC-14-11°
Pale Campanula Blue	XXIV	55*f	8.5	PB 7.5/ 6.0	U-16- 8°
Pale Cendre Green	VI	35f	7.5	GY 8.8/ 3.0	LLG-18- 8°
Pale Cerulean Blue	VIII	45d	10.	B 7.5/ 6.0	CCU-15-12°
*Pale Chalcedony Yellow	XVII	25′f	7.5	Y 8.5/ 4.0	YYL-18- 6°
Pale Cinnamon-Pink	XXIX	13″f	6.5	YR 8.0/ 3.0	SSO-18- 6°
Pale Congo Pink	XXVIII	7″f	10.	R 8.0/ 3.0	S-17- 7°
Pale Drab-Gray	XLVI	17″″f	10.	YR 8.5/ 0.5	SO-15- 1°
Pale Dull Glaucous-Blue	XLII	41‴f	2.5	B 7.5/ 2.0	TC-16- 4°
*Pale Dull Green-Yellow	XVII	27′f	2.5	GY 8.2/ 4.0	YL-18- 6°
Pale Ecru-Drab	XLVI	13″″f	5.	YR 8.0/ 1.0	SSO-15- 3°
Pale Flesh Color	XIV	7′f	2.5	YR 8.0/ 3.5	SSO-17- 7°
Pale Fluorite Green	XXXII	33″f	10.	GY 8.6/ 2.0	L-18- 6°
Pale Forget-me-not Blue	XXII	51*f	5.	PB 7.4/ 4.0	UUC-16- 8°
Pale Glass Green	XXXI	29″f	2.5	GY 8.4/ 3.0	LLY-18- 6°
Pale Glaucous-Blue	XXXIV	43″f	5.	B 8.0/ 3.0	TC-16- 4°
Pale Glaucous-Green	XXXIII	39″f	4.	G 8.4/ 2.0	GE-18- 6°
Pale Grayish Blue	XXI	49′f	5.	PB 7.6/ 4.0	UUC-16-11°
Pale Grayish Blue-Violet	XXXV	51″f	8.5	PB 7.6/ 4.0	UUV-16- 6°
Pale Grayish Vinaceous	XXXIX	9‴f	2.5	YR 8.0/ 1.5	S-16- 4°
Pale Grayish Violet-Blue	XXIV	53*d	8.5	PB 7.0/ 8.0	U-14-10°
*Pale Greenish Yellow	V	25d	10.	Y 8.0/ 7.0	YYL-18-10°
Pale Green-Blue Gray	XLVIII	45″″f	10.	B 6.5/ 2.0	UUC-14- 3°
Pale Green-Yellow	V	27f	1.	GY 8.5/ 5.0	YYL-19-10°
Pale Gull Gray	LIII	CG		N 7.8/	UV-16- 1°
Pale Hortense Violet	XI	61f	5.	P 7.5/ 6.0	VVU-16-12°
Pale King's Blue	XXII	47*f	10.	PB 8.5/ 4.0	UUC-16- 7°
Pale Laelia Pink	XXXVIII	67″f	2.5	RP 7.5/ 6.0	MR-15- 5°
Pale Lavender-Violet	XXV	61′f	5.	P 7.4/ 5.0	V-16- 9°
*Pale Lemon Yellow	IV	23b	8.5	Y 8.5/10.0	Y-18-11°
Pale Lilac	XXXVII	63″f	10.	P 7.8/ 4.0	VVM-16- 6°
Pale Lobelia Violet	XXXVII	61‴f	7.5	P 7.5/ 3.0	V-16- 6°
*Pale Lumiere Green	XVII	29′f	2.5	GY 8.0/ 4.5	L-18- 6°
Pale Mauve	XXV	63′f	6.	P 8.5/ 4.0	VVU-16-10°
Pale Mazarine Blue	IX	49f	5.	PB 7.5/ 4.0	UUC-17-12°

Ridgway Color Name	Ridgway Notation		Hamly Notation		Villalobos Notation
Pale Medici Blue	XLVIII	41''''f	10.	B 6.5/ 1.5	UUC-14- 3°
*Pale Methyl Blue	VIII	47d	10.	B 7.2/ 6.0	CCU-15-10°
Pale Mouse Gray	LI	15''''''d	7.5	YR 6.5/ 1.0	VM-13- 2°
Pale Neropalin Blue	XXII	49*f	2.5	PB 8.0/ 3.5	UUC-16- 6°
Pale Neutral Gray	LIII	NGd	N 5.9/		UV-13- 1°
Pale Niagara Green	XXXIII	41''f	7.5	BG 7.8/ 2.0	ET-18- 6°
Pale Nile Blue	XIX	41'f	7.5	BG 8.4/ 4.0	T-16- 7°
Pale Ochraceous-Buff	XV	15'f	1.	YR 8.2/ 5.0	OOS-18- 8°
Pale Ochraceous-Salmon	XV	13'f	1.	Y 8.5/ 4.0	OOS-18- 8°
Pale Olive-Buff	XL	21'''f	5.	Y 8.4/ 2.5	OY-18- 4°
Pale Olive-Gray	LI	23''''''f	7.5	Y 8.5/ 1.0	N-15/
Pale Olivine	XXXII	35''f	10.	GY 8.2/ 2.0	GGL-17- 4°
*Pale Orange-Yellow	III	17f	1.	Y 8.0/ 8.0	OOY-18-10°
Pale Payne's Gray	XLIX	49''''f	6.5	PB 6.4/ 2.0	U-14- 2°
Pale Persian Lilac	XXXVIII	69''f	5.	RP 7.4/ 8.0	MR-16- 6°
Pale Pinkish Buff	XXIX	17''f	10.	YR 8.2/ 2.5	OOS-18- 6°
Pale Pinkish Cinnamon	XXIX	15''f	8.5	YR 8.4/ 4.0	SO-18- 6°
Pale Purple-Drab	XLV	1''''d	7.5	RP 6.6/ 2.0	R-14- 3°
Pale Purplish Gray	LIII	67''''''d	10.	P 5.8/ 1.0	VVU-13- 1°
Pale Purplish Vinaceous	XXXIX	1'''f	10.	RP 8.0/ 4.5	R-16- 4°
Pale Quaker Drab	LI	1''''''d	7.5	RP 6.5/ 1.0	VVM-13- 2°
Pale Rhodonite Pink	XXXVIII	71''f	6.5	RP 7.6/ 7.0	R-16- 7°
Pale Rose-Purple	XXVI	67'f	8.5	P 6.5/ 7.0	VVM-16- 8°
Pale Rosolane Purple	XXVI	69'd	3.5	RP 6.4/12.0	M-15- 9°
Pale Russian Blue	XLII	45'''f	2.5	PB 8.5/ 4.0	UUC-15- 4°
Pale Salmon Color	XIV	9'f	7.5	YR 8.0/ 3.5	SSO-18- 8°
Pale Smoke Gray	XLVI	21''''f	5.	Y 7.5/ 1.5	OOS-15- 1°
Pale Soft Blue-Violet	XXIII	55'd	9.	PB 7.0/ 8.0	UUV-15-12°
Pale Sulphate Green	XIX	39'd	2.5	BG 7.0/ 4.0	E-15- 9°
Pale Tiber Green	XVIII	33'f	7.5	GY 8.5/ 4.0	LG-17- 7°
Pale Turquoise Green	VII	41f	5.	BG 8.0/ 4.0	T-16-11°
Pale Turtle Green	XXXII	31''f	6.5	GY 8.2/ 3.0	LLY-18- 4°
Pale Varley's Gray	XLIX	57''''d	2.5	P 6.0/ 3.0	UV-12- 2°
Pale Verbena Violet	XXXVI	55''f	10.	PB 7.6/ 6.0	UV-15- 8°
Pale Veronese Green	XVIII	31'f	5.	GY 8.2/ 3.0	LLG-18- 6°
*Pale Vinaceous	XXVII	1''f	2.5	R 7.8/ 5.0	R-16- 6°
Pale Vinaceous-Drab	XLV	5''''d	2.5	R 6.5/ 1.5	RS-14- 3°
Pale Vinaceous-Fawn	XL	13'''f	5.	YR 8.0/ 1.5	SSO-16- 3°
Pale Vinaceous-Lilac	XLIV	69'''f	2.5	RP 7.8/ 5.0	MMV-16- 5°
Pale Vinaceous-Pink	XXVIII	9''f	2.5	YR 8.0/ 2.5	SSO-17- 7°
Pale Violet	X	59d	9.	PB 6.5/ 8.0	UUV-14-12°
Pale Violet-Blue	IX	53d	6.5	PB 7.0/10.0	U-14-12°
Pale Violet-Gray	LII	59''''''d	2.5	P 6.4/ 0.5	UV-13- 2°
Pale Violet-Plumbeous	XLIX	53''''f	8.5	PB 7.0/ 3.0	UUV-15- 3°
Pale Viridine Yellow	V	29f	1.5	GY 8.4/ 5.0	YL-19-12°
Pale Windsor Blue	XXXV	49''''d	7.	PB 7.0/ 7.0	U-14- 7°
Pale Wistaria Blue	XXIII	57'f	10.	PB 7.5/ 6.0	UUV-16-12°
Pale Wistaria Violet	XXIII	59'f	2.5	P 7.2/ 5.0	UUV-16-12°

204

Ridgway Color Name	Ridgway Notation		Hamly Notation		Villalobos Notation
Pale Yellow-Green	VI	31f	5.	GY 8.4/ 5.0	LLG-18- 8°
Pale Yellow-Orange	III	15f	1.5	YR 8.4/ 6.5	O-18-10°
Pallid Blue-Violet	X	55f	8.5	PB 7.5/ 6.0	UUV-16-12°
Pallid Bluish Violet	X	57f	9.	PB 7.5/ 6.0	UUV-16-12°
Pallid Brownish Drab	XLV	9''''f	5.	YR 8.0/ 1.0	RS-16- 2°
Pallid Grayish Violet- Blue	XXIV	53*f	7.5	PB 7.2/ 5.0	U-15- 8°
Pallid Methyl Blue	VII	47f	2.5	PB 8.0/ 6.0	CCU-17-11°
Pallid Mouse Gray	LI	15''''f	5.	YR 8.0/ 0.5	MMV-15- 1°
Pallid Neutral Gray	LIII	NGf	N 7.4/		UV-15- 1°
Pallid Purple-Drab	XLV	1''''f	7.5	RP 7.2/ 1.5	MMV-15- 3°
Pallid Purplish Gray	LIII	67'''''f	10.	P 8.0/ 1.0	UV-16- 2°
Pallid Quaker Drab	LI	1'''''f	7.5	RP 8.0/ 1.0	VVM-15- 3°
Pallid Soft Blue-Violet	XXIII	55'f	9.	PB 7.5/ 6.0	UUV-16-12°
Pallid Vinaceous-Drab	XLV	5''''f	7.5	RP 8.0/ 1.5	RS-16- 2°
Pallid Violet	X	59f	9.	PB 7.5/ 5.0	UUV-15-12°
Pallid Violet-Blue	IX	53f	6.5	PB 7.5/ 6.0	U-16-12°
Pansy Purple	XII	69k	7.5	RP 3.2/ 6.0	R- 5- 7°
Pansy Violet	XI	63i	7.5	P 4.0/12.0	VVM- 5-10°
Paris Blue	VIII	47k	5.	PB 3.0/10.0	CU- 5-12°
Paris Green	XVIII	35'b	2.5	G 6.8/ 6.5	G-13- 8°
Parrot Green	VI	31k	5.	GY 5.5/ 5.5	LLY- 8- 5°
Parula Blue	XLII	45'''	2.5	PB 5.5/ 4.0	UUC-11- 6°
Patent Blue	VIII	43k	7.5	B 3.0/ 6.0	CCU- 4-10°
Payne's Gray	XLIX	49''''	6.5	PB 4.4/ 2.5	UUV-10- 3°
Peach Red	I	5b	7.5	R 5.5/ 9.4	SSO-14-12°
Peacock Blue	VIII	43i	5.	B 3.6/ 8.5	C- 6-11°
Peacock Green	VI	35i	7.5	GY 6.2/ 5.0	GGL-11- 5°
Pea Green	XLVII	29''''b	10.	GY 6.4/ 1.0	L-12- 2°
Pearl Blue	XXXV	49''f	7.5	PB 7.5/ 4.0	UUV-16- 4°
Pearl Gray	LII	35'''''f	2.5	B 8.0/ 0.2	CCU-16- 1°
*Pecan Brown	XXVIII	11''i	5.	YR 5.0/ 4.5	SO-10- 5°
Perilla Purple	XXXVII	65''d	7.5	RP 3.4/ 4.5	MR- 5- 4°
Persian Blue	XX	45'f	2.5	PB 7.5/ 4.0	CU-16-10°
Persian Lilac	XXXVIII	69''d	5.	RP 6.2/ 7.0	MR-13- 7°
Petunia Violet	XXV	65'i	5.	P 3.4/ 8.0	VVM- 8- 8°
*Phenyl Blue	IX	53	7.5	PB 3.5/16.0	U- 6-13°
				or	
			8.	PB 3.5/17.0	
Phlox Pink	XI	65f	7.5	P 7.0/ 6.0	V-16-12°
*Phlox Purple	XI	65b	10.	P 5.0/11.0	VM-11-11°
*Picric Yellow	IV	23d	8.5	Y 8.5/ 9.5	Y-18/19-11°
Pinard Yellow	IV	21d	5.	Y 8.4/ 8.0	YYO-18-11°
Pinkish Buff	XXIX	17''d	1.	YR 8.0/ 4.0	O-17- 6°
Pinkish Cinnamon	XXIX	15''b	7.5	YR 6.6/ 6.0	OOS-14- 7°
Pinkish Vinaceous	XXVII	5''d	7.5	R 7.5/ 5.5	S-14- 7°
Pistachio Green	XLI	33'''	10.	GY 5.5/ 3.0	GGL-12- 4°
Pleroma Violet	XXV	61'	2.5	P 4.5/11.0	VVU- 8- 8°
Plumbago Blue	XLIII	53'''f	10.	PB 7.2/ 3.0	UV-15- 5°

205

Ridgway Color Name	Ridgway Notation		Hamly Notation			Villalobos Notation
Plumbago Slate	L	61''''i	7.5	P 4.0/	1.0	UV- 7- 4°
Plumbago Gray	L	61''''d	10.	P 5.4/	1.0	VVU-12- 2°
Plumbeous	LII	49'''''b	7.5	PB 5.6/	1.5	UUV-12- 2°
Plumbeous-Black	LII	49'''''m	7.5	PB 2.2/	0.5	U- 2- 3°
Plum Purple	XXIV	57*m	10.	PB 2.4/	4.0	UUV- 3- 9°
Pois Green	XLI	29'''i	5.	GY 5.0/	2.5	LLY- 8- 3°
Pomegranate Purple	XII	71i	5.	RP 3.8/	7.0	R- 8-11°
Porcelain Blue	XXXIV	43''	2.5	B 5.6/	3.5	C-10- 3°
Porcelain Green	XXXIII	39''i	2.5	BG 4.8/	3.0	ET- 9- 4°
Pompeian Red	XIII	3'i	5.	R 4.2/	8.0	S- 8- 7°
*Primrose Yellow	XXX	23''d	7.5	Y 8.2/	6.5	Y-18- 8°
Primuline Yellow	XVI	19'	3.5	Y 7.5/	10.0	OY-16-11°
*Prout's Brown	XV	15'm	7.5	YR 3.5/	3.0	OOS- 5- 5°
Prune Purple	XI	63m	7.5	P 2.2/	6.0	V- 3- 8°
*Prussian Blue	IX	49m	3.5	PB 3.2/	9.0	UUC- 4-10°
Prussian Green	XIX	41'k	7.5	BG 3.4/	4.0	T- 6- 6°
Prussian Red	XXVII	5''k	7.5	R 4.0/	3.5	S- 7- 5°
Puritan Gray	XLVII	33''''f	7.5	G 7.0/	1.5	CCU-14- 1°
Purple (true)	XI	65	10.	P 4.0/	14.0	VM- 8-12°
				or		
			9.5	P 4.2/	17.0	
Purple-Drab	XLV	1''''	8.	RP 4.6/	2.0	R-10- 3°
Purplish Gray	LIII	67'''''	10.	P 4.0/	1.5	VVU- 9- 1°
Purplish Lilac	XXXVII	65''d	2.5	RP 6.5/	8.0	VM-13- 5°
Purplish Vinaceous	XXXIX	1'''b	2.5	R 6.2/	4.5	R-12- 6°
Pyrite Yellow	IV	23i	7.5	Y 6.0/	7.0	Y-13-10°
Quaker Drab	LI	1'''''	5.	RP 4.6/	0.5	MR-10- 1°
Rainette Green	XXXI	27''i	2.5	GY 5.5/	4.0	YYL-11- 4°
Ramier Blue	XLIII	57'''	2.5	P 4.5/	3.0	VVU-10- 4°
Raisin Black	XLIV	65'''m	2.5	RP 2.6/	2.0	MMV- 4- 3°
Raisin Purple	XI	65k	10.	P 2.8/	6.0	VM- 4-10°
Raw Sienna	III	17i	7.5	YR 5.5/	9.0	O-12-12°
Raw Umber	III	17m	7.5	YR 3.5/	3.0	O- 5- 5°
Reed Yellow	XXX	23''b	7.5	Y 7.8/	6.0	YYO-16- 6°
Rejane Green	XXXIII	37''b	2.5	G 6.5/	4.5	GGL-14- 5°
Rhodamine Purple	XII	67	1.	P 4.5/	14.0	M-10-12°
				or		
			2.	RP 4.3/	17.5	
Rhodonite Pink	XXXVIII	71''d	7.5	RP 6.0/	7.0	R-14- 8°
Rinnemann's Green	XVIII	31'i	7.5	GY 5.5/	4.5	LLG-10- 5°
*Rivage Green	XVIII	31'b	6.5	GY 7.5/	6.5	LLG-15- 9°
Rocellin Purple	XXXVIII	71''b	7.5	RP 5.8/	9.0	R-12- 7°
*Roman Green	XVI	23'm	7.5	Y 4.0/	3.5	Y- 4- 5°
Rood's Blue	IX	49k	6.5	PB 3.5/	14.0	UUC- 5-12°
Rood's Brown	XXVIII	11''k	4.	YR 4.5/	3.0	SO- 8- 4°
Rood's Lavender	XLIX	57''''f	2.5	P 6.5/	2.0	UV-14- 3°
Rood's Violet	XI	65i	10.	P 3.0/	9.0	MMV- 4-11°
Rose Color	XII	71b	5.	RP 5.5/	14.0	MR-12-11°
*Rose Doree	I	3b	6.	R 6.0/	12.0	RS-13-12°

206

Ridgway Color Name	Ridgway Notation		Hamly Notation		Villalobos Notation
*Rose Pink	XII	71f	5.	RP 8.0/ 8.0	M-17-12°
Rose Purple	XXVI	67'd	8.5	P 6.5/10.0	VVM-14-11°
*Rose Red	XII	71	7.5	RP 4.8/14.0	R-10-12°
Rosolane Pink	XXVI	69'f	5.	RP 8.0/ 9.0	MR-16- 7°
*Rosolane Purple	XXVI	69'	5.	RP 4.4/12.0	MR- 9- 8°
Roslyn Blue	X	57k	9.	PB 3.0/ 8.0	UUV- 3-12°
*Royal Purple	X	59i	1.	P 3.5/ 8.0	UV- 5-11°
*Rufous	XIV	9'	10.	R 5.2/10.0	SO-13- 9°
Russet	XV	13'k	5.	YR 3.8/ 4.0	SO- 8- 4°
*Russet-Vinaceous	XXXIX	9'''	10.	R 5.0/ 3.0	SSO- 9- 5°
Russian Blue	XLII	45'''d	2.5	PB 7.5/ 3.0	UUC-14- 5°
Russian Green	XLII	37'''i	2.5	G 4.0/ 1.5	G- 9- 3°
Saccardo's Olive	XVI	19'm	2.5	Y 4.5/ 4.0	OY- 7- 6°
Saccardo's Slate	XLVIII	41''''k	7.5	B 2.8/ 1.0	O- 8- 3°
Saccardo's Umber	XXIX	17''k	1.	Y 4.0/ 2.0	UUC- 5- 3°
Saccardo's Violet	XXXVII	61''	6.	P 4.8/ 9.0	V- 8- 8°
Safrano Pink	II	7f	10.	R 8.0/ 8.0	SSO-18-12°
Sage Green	XLVII	29''''	10.	GY 5.6/ 1.5	LG-10- 2°
Sailor Blue	XXI	53'k	8.5	PB 3.5/10.0	U- 5- 9°
*Salmon-Buff	XIV	11'd	7.5	YR 7.8/ 6.0	SO-17- 8°
Salmon Color	XIV	9'd	5.	YR 7.5/ 6.0	SO-16- 9°
Salmon-Orange	II	11b	2.5	YR 7.0/11.0	OOS-15-12°
Salvia Blue	IX	49b	7.5	PB 5.0/16.0	UUC-14-13°
*Sanford's Brown	II	11k	2.5	YR 4.2/ 8.0	SO- 8-11°
*Sayal Brown	XXIX	15''i	8.5	YR 5.2/ 5.0	OOS-10- 5°
Scarlet	I	5	7.5	R 5.0/12.0	SSO-10-12°
*Scarlet-Red	I	3	5.	R 4.4/16.0	RS-10-12°
*Scheele's Green	VI	33i	7.5	GY 6.5/ 7.0	L-13-10°
*Schoenfeld's Purple	XXVI	69'i	5.	RP 3.2/10.0	MR- 8- 7°
Seafoam Green	XXXI	27''f	1.	GY 8.2/ 4.0	YL-19- 6°
Seafoam Yellow	XXXI	25''f	7.5	Y 8.4/ 5.0	YYL-19- 6°
Sea Green	XIX	41'i	7.5	BG 4.5/ 6.0	T- 8- 7°
*Seal Brown	XXXIX	5'''m	7.5	R 3.2/ 1.5	RS- 5- 2°
Seashell Pink	XIV	11'f	7.5	YR 8.0/ 3.0	SO-18- 6°
Sepia	XXIX	17''m	1.	Y 3.4/ 2.0	OOS- 5- 4°
Serpentine Green	XVI	23'k	7.5	Y 4.8/ 4.0	Y- 8- 8°
Shamrock Green	XXXII	33''i	1.	G 5.4/ 4.0	LG- 9- 4°
Shell Pink	XXVIII	11''f	2.5	YR 8.0/ 2.5	SSO-17- 8°
Shrimp Pink	I	5f	10.	R 8.0/ 6.0	S-17-11°
Skobeloff Green	VII	39	2.5	BG 4.8/10.0	ET- 9-11°
Sky Blue	XX	47'd	2.5	PB 7.0/ 7.0	CU-15-11°
Sky Gray	XXXIV	45''f	2.5	PB 8.5/ 2.0	UUC-16- 5°
Slate-Black	LIII	CG	N 2.4/		UUV- 2- 1°
Slate-Blue	XLIII	49'''i	7.5	PB 4.0/ 3.5	U- 8- 5°
Slate Color	LIII	CGk	N 3.4/		U- 5- 2°
Slate-Gray	LIII	CGi	N 3.5/		U- 9- 1°
Slate-Olive	XLVII	29''''i	3.5	G 5.0/ 1.0	GGL- 8- 1°
Slate-Purple	XLIV	65'''i	2.5	RP 4.0/ 3.0	MMV- 7- 4°
Slate-Violet (1)	XLIII	57'''i	1.	P 3.4/ 1.5	UV- 6- 5°

207

Ridgway Color Name	Ridgway Notation		Hamly Notation		Villalobos Notation
Slate-Violet (2)	XLIV	61'''	7.5	P 4.5/ 3.0	V- 8- 5°
Smalt Blue	IX	53i	8.5	PB 3.2/14.0	U- 6-12°
Smoke Gray	XLVI	21''''d	5.	Y 6.8/ 2.0	O-13- 2°
*Snuff Brown	XXIX	15''k	9.	YR 4.5/ 3.0	OOS- 7- 5°
Soft Blue-Violet	XXIII	55'	10.	PB 4.8/12.0	UUV- 9-10°
Soft Bluish Violet	XXIII	57'	1.	P 4.2/12.0	UV- 9- 1°
Sooty Black	LI	1'''''m	5.	RP 3.0/ 0.5	VVU- 3- 1°
*Sorghum Brown	XXXIX	9'''i	10.	R 3.5/ 2.0	SO- 9- 3°
Sorrento Green	VII	41k	9.	BG 3.8/ 6.5	T- 5- 8°
Spectrum Blue	IX	49	5.	PB 4.0/16.0	UUC- 8-13°
Spectrum Red	I	1	5.	R 4.0/15.0	RS- 9-12°
Spectrum Violet	X	59	10.	PB 3.5/10.0	UV- 6-12°
				or	
			10.	PB 3.3/14.0	
*Spinach Green	V	29m	5.	GY 4.6/ 4.0	L- 8- 4°
*Spinel Pink	XXVI	71'b	5.	RP 5.2/12.0	MR-13-11°
*Spinel Red	XXVI	71'	8.5	RP 5.0/12.0	R-10-10°
Squill Blue	XX	45'b	2.5	PB 6.2/ 8.0	CU-13-10°
Stone Green	XLII	37'''	2.5	G 5.5/ 3.0	G-10- 4°
Storm Gray	LII	35'''''	5.	G 5.4/ 1.0	E-10- 1°
Strawberry Pink	I	5d	7.5	R 6.4/10.0	SSO-15-12°
Straw Yellow	XVI	21'd	6.	Y 8.4/ 6.0	YYO-18- 8°
*Strontian Yellow	XVI	23'	7.5	Y 7.5/ 9.0	Y-16-11°
*Sudan Brown	III	15k	7.5	YR 4.5/ 6.0	O- 8-12°
Sulphate Green	XIX	39'	10.	G 5.5/ 6.0	ET-10- 8°
Sulphine Yellow	IV	21i	5.	Y 5.8/ 6.0	YYO-14-11°
Sulphur Yellow	V	25f	7.5	Y 8.4/ 5.0	YYL-19- 9°
Taupe Brown	XLIV	69'''m	5.	RP 2.6/ 2.0	MR- 4- 3°
Tawny	XV	13'i	5.	YR 5.0/ 6.5	OOS-10- 7°
*Tawny-Olive	XXIX	17''i	1.	Y 5.2/ 4.5	O-11- 4°
Tea Green	XLVII	25''''b	5.	GY 6.2/ 2.0	Y-11- 1°
Terra Cotta	XXVIII	7''	10.	R 5.4/ 6.5	SSO-10- 7°
Terre Verte	XXXIII	41''i	4.	BG 4.5/ 2.0	E- 8- 4°
Testaceous	XXVIII	9''	2.5	YR 5.0/ 5.0	SO-11- 6°
*Thulite Pink	XXVI	71'd	5.	RP 6.6/11.0	R-16-10°
Tiber Green	XVIII	33'd	10.	GY 8.0/ 6.0	LG-16- 9°
Tilleul Buff	XL	17'''f	7.5	YR 8.4/ 2.0	OOS-17- 4°
Tourmaline Pink	XXXVIII	67''b	2.5	RP 5.5/11.0	MR-11- 6°
Turquoise Green	VII	41d	5.	BG 7.0/ 6.0	ET-15-12°
Turtle Green	XXXII	31''b	7.5	GY 7.0/ 4.5	L-14- 5°
Tyrian Blue	XXXIV	47''i	5.	PB 4.2/ 3.0	U- 9- 7°
*Tyrian Pink	XII	69b	2.5	RP 5.4/12.0	MR-12-11°
*Tyrian Rose	XII	69	5.	RP 4.5/14.0	R- 9-12°
				or	
			8.	RP 4.3/15.0	
Tyrolite Green	VII	39b	10.	G 5.6/ 6.0	ET-12-10°
Ultramarine Ash	XXII	49*	6.5	PB 5.0/12.0	UUC- 9-11°
*Ultramarine Blue	IX	49i	6.5	PB 3.5/16.0	UUC- 7-13°
Urania Blue	XXIV	53*m	10.	PB 2.5/ 3.0	UUV- 4- 6°

208

Ridgway Color Name	Ridgway Notation		Hamly Notation	Villalobos Notation
Vanderpoel's Blue	XX	47i	3.5 PB 4.6/ 9.0	UUC- 9-11°
*Vanderpoel's Green	VI	33b	7.5 GY 7.5/10.0	GGL-15-11°
Vanderpoel's Violet	XXXVI	55''	1. P 4.8/ 6.0	UV- 8- 8°
*Vandyke Brown	XXVIII	11''m	5. YR 3.5/ 2.5	SSO- 5- 3°
*Vandyke Red	XIII	1'k	2.5 R 3.8/ 5.0	RS- 7- 5°
Variscite Green	XIX	37'd	3.5 G 7.6/ 4.0	G-15- 9°
Varley's Gray	XLIX	57''''	2.5 P 4.6/ 2.5	UV- 9- 4°
Varley's Green	XVIII	31'm	9. GY 4.0/ 3.0	LG- 5- 2°
Venetian Blue	XXII	47*	5. PB 4.5/10.0	UUC-10-11°
Venetian Pink	XIII	1'f	7.5 R 8.0/ 6.0	RS-17- 7°
*Venice Green	VII	41b	4. BG 6.0/ 8.0	ET-13-11°
Verbena Violet	XXXVI	55''d	10. PB 6.5/ 4.0	UV-13- 6°
*Verdigris Green	XIX	37'	3.5 G 6.0/ 5.0	E-10- 8°
Vernonia Purple	XXXVIII	69''i	7.5 RP 4.2/ 6.5	MR- 7- 5°
Verona Brown	XXIX	13''k	5. YR 4.2/ 3.0	OOS- 7- 5°
*Veronese Green	XVIII	31'd	7.5 GY 7.5/ 6.0	LLG-16- 8°
Vetiver Green	XLVII	25''''	5. GY 5.6/ 3.0	Y- 9- 3°
*Victoria Lake	I	1m	5. R 2.0/ 4.0	RS- 4- 6°
Vinaceous	XXVII	1''d	2.5 R 6.5/ 7.0	R-14- 8°
Vinaceous-Brown	XXXIX	5'''i	7.5 R 4.8/ 2.5	S- 8- 4°
Vinaceous-Buff	XL	17'''d	6.5 YR 8.0/ 3.0	OOS-15- 5°
*Vinaceous-Cinnamon	XXIX	13''b	7. YR 7.2/ 5.0	OOS-14- 7°
Vinaceous-Drab	XLV	5''''	2.5 R 5.0/ 2.0	RS-10- 3°
Vinaceous-Fawn	XL	13'''b	2.5 YR 6.8/ 3.0	SO-13- 4°
Vinaceous-Gray	L	69''''d	7.5 P 5.5/ 2.0	VVM-12- 3°
Vinaceous-Lavender	XLIV	65'''f	10. RP 7.6/ 6.0	VVM-15- 5°
Vinaceous-Lilac	XLIV	69'''b	5. RP 5.5/ 6.0	MR-11- 5°
Vinaceous-Pink	XXVIII	9''d	2.5 YR 6.0/ 5.0	SO-17- 7°
Vinaceous-Purple (1)	XXXVIII	67''i	3.5 RP 4.5/ 7.0	M-11- 5°
Vinaceous-Purple (2)	XLIV	65'''	10. P 3.8/ 3.5	MMV- 9- 4°
*Vinaceous-Rufous	XIV	7'i	10. R 4.5/ 9.0	SO- 9-10°
*Vinaceous-Russet	XXVIII	7''i	2.5 YR 4.5/ 4.0	SO-10- 4°
Vinaceous-Slate	L	69''''i	7.5 P 3.8/ 2.0	VVM- 6- 3°
Vinaceous-Tawny	XXVIII	11''	2.5 YR 5.6/ 6.0	SO-12- 7°
Violet Carmine	XII	69m	7.5 RP 3.0/ 4.0	R- 4- 5°
Violet-Gray	LII	59'''''	2.5 P 5.4/ 1.0	UV- 9- 2°
Violet-Plumbeous	XLIX	53''''b	8.5 PB 5.2/ 4.0	UUV-11- 4°
Violet-Purple	XI	63	6.5 P 5.0/14.0	VVM- 8-12°
			or	
			5. P 4.1/15.0	
Violet-Slate	XLIX	53''''i	8.5 PB 3.6/ 2.5	UV- 7- 3°
*Violet-Ultramarine	X	57i	9. PB 3.4/12.0	UUV- 4-12°
*Viridian Green	VII	37i	5. G 5.5/ 8.0	E- 8- 8°
*Viridine Green	VI	33d	6.5 GY 8.0/ 8.0	LLG-17-10°
*Viridine Yellow	V	29b	1. GY 7.8/ 8.0	L-17-12°
*Vivid Green	VII	37	4. G 6.0/10.0	GE-11-11°
Wall Green	VII	39k	5. BG 3.8/ 6.0	ET- 6- 8°
*Walnut Brown	XXVIII	9''k	2.5 YR 4.0/ 3.0	SO- 8- 4°
*Warbler Green	IV	23k	7.5 Y 5.0/ 5.0	V- 9-10°

Ridgway Color Name	Ridgway Notation		Hamly Notation	Villalobos Notation
*Warm Blackish Brown	XXXIX	1'''m	2.5 R 3.4/ 1.5	R- 5- 2°
*Warm Buff	XV	17'd	1. Y 7.8/ 6.0	O-17- 7°
*Warm Sepia	XXIX	13''m	5. YR 3.8/ 2.0	SO- 6- 3°
Water Green	XLI	25'''d	2.5 GY 7.6/ 2.0	YL-16- 2°
Wax Yellow	XVI	21'	5. Y 7.0/ 9.0	YYO-16-11°
Wedgwood Blue	XXI	51'f	7. PB 7.5/ 8.0	U-16-12°
*White	LIII	NGo	N 9.5/	N-20/
Windsor Blue	XXXV	49''i	8. PB 4.0/ 5.5	U- 8- 8°
Winter Green	XVIII	33'i	2.5 G 4.8/ 4.0	GGL- 9- 5°
*Wistaria Blue	XXIII	57'b	9. PB 5.8/10.0	UUV-13-11°
Wistaria Violet	XXIII	59'b	2.5 P 6.0/12.0	UUV-13-12°
*Wood Brown	XL	17'''	9. YR 6.0/ 3.0	OOS-11- 4°
*Xanthine Orange	III	13i	5. YR 5.2/12.0	OOS-11-12°
*Yale Blue	XX	47'b	3.5 PB 6.0/ 8.0	UUC-13-11°
Yellow-Green	VI	31	5. GY 8.0/10.0	L-16-10°
			or	
			4.5 GY 7.2/11.0	
Yellowish Citrine	XVI	23'i	7.5 Y 6.0/ 6.0	Y-11-10°
Yellowish Glaucous	XLI	25'''f	2.5 GY 8.4/ 2.0	LLY-16- 2°
Yellowish Oil Green	V	25k	10. Y 5.0/ 6.0	YYL- 9- 3°
*Yellowish Olive	XXX	23''k	7.5 Y 4.5/ 4.0	YYO- 7- 4°
Yellow Ocher	XV	17'	10. YR 5.8/ 7.0	OOY-14-11°
Yew Green	XXXI	27''m	5. GY 3.4/ 3.5	YL- 5- 2°
Yvette Violet	XXXVI	55''k	1. P 3.5/ 4.0	VVU- 5- 6°
Zinc Green	XIX	37'i	5. G 5.4/ 4.0	GE- 9- 5°
Zinc Orange	XV	13'	5. YR 6.5/10.0	OOS-14-10°

210

APPENDIX B
NOMENCLATURE
CORRELATION

Appendix B correlates the nomenclature used by Ridgway in his descriptions of birds in Bulletin 50 with modern nomenclature. The citations in the Correlated Notes quote the scientific nomenclature verbatim, as it occurs in Bulletin 50. However, much of Ridgway's nomenclature has become obsolete due to changes made in the scientific names of birds by taxonomists since his writings. Appendix B lists the name of bird cited, with the volume and page number of its occurrence in Bulletin 50. If there has been no change between the old and the new scientific name, the right-hand column is blank. If there is a more modern name, the new nomenclature is entered in that column.

Appendix B is limited to the names of birds cited in the Correlated Notes. No attempt has been made to develop a similar correlation for Ridgway's colloquial bird names.

To accomplish this nomenclature correlation, an arduous task, I enlisted the aid of Charles O'Brien, Curator Emeritus of the Ornithology Department of The American Museum of Natural History. The citations in the Correlated Notes were submitted to him for review and modernization. O'Brien has been a lifelong witness to the many changes that have developed, year after year, since Ridgway's time. It would be difficult, if not impossible, to find an ornithologist so well qualified to make these correlations. I am very grateful indeed to him for his willingness to accomplish the task, and for the skill with which he did it.

Vol. XI	Ridgway Nomenclature	Recent Nomenclature
743	*Falco sparverius sparveriodes*	*Falco sparverius sparveroides*
723	*Falco sparverius sparverius*	
675	*Falco a. albigularis*	*Falco r. rufigularis*
674	*Falco deiroleucus*	
555	*Herpetotheres c. cachinnans*	
483	*Heliaeetus l. leucocephalus*	
444	*Spizaëtus ornatus vicarius*	*Spizaetus ornatus vicarius*
338	*Buteo magnirostris griseocauda*	
280	*Buteo l. lineatus*	
246	*Buteo jamaicensis colurus*	
235	*Buteo jamaicensis borealis*	
226	*Buteo regalis*	
205	*Heterospizias m. meridionalis*	
195	*Accipiter striatus suttoni*	
168	*Accipiter b. bicolor*	
132	*Rostrhamus s. sociabilis*	
125	*Ictinia plumbea*	
117	*Harpagus bidentatus fasciatus*	
103	*Chondrohierax u. uncinatus*	
12	*Sarcorhamphus papa*	

Vol. X		
345	*Colinus virginianus ridgwayi*	
341	*Colinus virginianus salvini*	
338	*Colinus virginianus insignis*	
313	*Colinus v. virginianus*	
212	*Tympanuchus cupido pinnatus*	
156	*Bonasa u. umbellus*	
49	*Ortalis wagleri griseiceps*	
48	*Ortalis w. wagleri*	
46	*Ortalis ruficauda*	
42	*Ortalis garrula cinereiceps*	
38	*Ortalis vetula pallidiventris*	
23	*Penelope p. purpurascens*	

Vol. IX		
169	*Laterallis r. ruber*	
168	*Laterallis ruber tamaulipensus*	
124	*Aramides axillaris*	
120	*Aramides c. cajanea*	
115	*Aramides cajanea mexicana*	
108	*Amaurolimnas c. concolor*	

Vol. VIII		
752	*Endomychura hypoleuca*	
721	*Uria t. troille*	*Uria a. aalge*
709	*Plautus alle*	*Plautus a. alle*
649	*Hydrocoloeus minutus*	*Larus minutus*
623	*Larus delawarensis*	
612	*Larus argentatus*	
597	*Larus glaucescens*	

212

Vol. VIII	Ridgway Nomenclature	Recent Nomenclature
585	*Larus hyperboreus*	
479	*Gelochelidon nilotica*	
437	*Recurvirostra americana*	
431	*Steganopus tricolor*	
424	*Lobipes lobatus*	
418	*Phalaropus fulicarius*	
412	*Mesoscolopax borealis*	*Numenius borealis* (extinct?)
397	*Phaeopus p. phaeopus*	*Numenius p. phaeopus*
353	*Tringa ocrophus*	
303	*Eurynorhynchus pygmeus*	
291	*Pisobia ruficollis*	*Calidris (Erolia) ruficollis*
250	*Erolia ferruginea*	*Calidris (Erolia) ferruginea*
244	*Arquatella p. ptilocnemis*	*Calidris (Erolia) p. ptilocnemis*
232	*Canutus canutus*	*Calidris c. canutus*
192	*Vetola haemastica*	*Limosa haemastica*
184	*Vetola fedoa*	*Limosa fedoa*
140	*Charadrius collaris*	
120	*Charadrius hiaticula*	*Charadrius h. hiaticula*
113	*Pagolla wilsonia cinnamomina*	*Charadrius wilsonia cinnamominus*
99	*Oxyechus v. vociferus*	*Charadrius v. vociferus*
55	*Arenaria melanocephala*	
40	*Haematopus bachmani*	
32	*Haematopus p. palliatus*	
9	*Jacana s. spinosa*	

Vol. VII

484	*Oreopeleia violacea albiventer*	
478	*Oreopeleia montana*	*Oreopeleia m. montana*
460	*Leptotila rufinucha*	*Leptotila cassini rufinucha*
456	*Leptotila c. cassini*	
451	*Leptotila f. fulviventris*	*Leptotila verreauxi fulviventris*
446	*Leptotila v. verreauxi*	
443	*Leptotila collaris*	*Leptotila jamaicensis collaris*
424	*Chaemepelia r. rufipennis*	*Columbigallina talpacoti rufipennis*
398	*Chaemepelia p. passerina*	*Columbigallina p. passerina*
390	*Scardafella inca*	
370	*Zenaida ruficauda vinaceo-rufa*	*Zenaida auriculata vinaceorufa*
363	*Zenaida aurita*	*Zenaida a. aurita*
361	*Zenaida zenaida lucida*	*Zenaida aurita zenaida*
357	*Zenaida z. zenaida*	*Zenaida aurita zenaida*
353	*Zenaidura yucatanensis*	Hybrid ♀ *Zenaida aurita yucatanensis* ♂ *Zenaidura macroura marginella*
352	*Zenaidura graysoni*	
351	*Zenaidura macroura tresmariae*	
341	*Zenaidura m. macroura*	
326	*Oenoenas s. subvinacea*	*Columba s. subvinacea*
321	*Crossophthalmus gymnophthalmus*	*Columba corensis*
316	*Lepidoenas speciosa*	*Columba speciosa*
305	*Chloroenas rufina pallidicrissa*	*Columba rufina pallidicrissa*

213

Vol. VII	Ridgway Nomenclature	Recent Nomenclature
292	*Chloroenas albilinea crissalis*	*Columba fasciata crissalis*
270	*Amazona l. leucocephala*	
266	*Amazona ventralis*	
262	*Amazona agilis*	
260	*Amazona xantholora*	
254	*Amazona a. albifrons*	
251	*Amazona b. barbadensis*	
247	*Amazona o. oratrix*	*Amazona ochrocephala oratrix*
242	*Amazona viridigenalis*	
241	*Amazona farinosa guatemalae*	
239	*Amazona farinosa inornata*	
234	*Amazona a. autumnalis*	
231	*Amazona auropalliata*	*Amazona ochrocephala auropalliata*
229	*Amazona arausiaca*	
227	*Amazona versicolor*	
225	*Amazona guildingii*	
223	*Amazona imperialis*	
210	*Pionus menstruus*	
203	*Pyrilia h. haematotis*	*Pionopsitta h. haematotis*
200	*Urochroma costaricensis*	*Touit dilectissima costaricensis*
199	*Urochroma dilectissima*	*Touit d. dilectissima*
194	*Psittacula insularis*	*Forpus cyanopygius insularis*
192	*Psittacula c. cyanopygia*	*Forpus c. cyanopygius*
183	*Brotogeris jugularis*	*Brotogeris j. jugularis*
179	*Bolborhynchus l. lineola*	
176	*Pyrrhura h. hoffmanni*	
174	*Eupsittula nana*	*Aratinga nana*
172	*Eupsittula astec*	*Aratinga a. astec*
169	*Eupsittula canicularis*	*Aratinga c. canicularis*
168	*Eupsittula ocularis*	*Aratinga pertinax ocularis*
164	*Eupsittula p. pertinax*	*Aratinga p. pertinax*
160	*Aratinga euops*	
159	*Aratinga brevipes*	*Aratinga holochlora brevipes*
157	*Aratinga h. holochlora*	
156	*Aratinga rubritorquis*	*Aratinga holochlora rubritorquis*
153	*Aratinga c. chloroptera*	
152	*Aratinga finschi*	
138	*Ara severa*	*Ara s. severa*
136	*Ara tricolor*	(extinct)
134	*Ara ambigua*	*Ara a. ambigua*
132	*Ara militaris mexicana*	
128	*Ara macao*	
126	*Ara chloroptera*	
122	*Ara ararauna*	
84	*Neomorphus salvini*	*Neomorphus geoffroyi salvini*
76	*Geococcyx californianus*	
71	*Morococcyx e. erythropygus*	
67	*Tapera naevia excellens*	
64	*Saurothera dominicensis*	*Saurothera vetula longirostris*

214

Vol. VII	Ridgway Nomenclature	Recent Nomenclature
62	*Saurothera vieilloti*	*Saurothera vetula vieilloti*
56	*Hyetornis rufigularis*	*Piaya rufigularis*
55	*Hyetornis pluvialis*	*Piaya pluvialis*
51	*Piaya mexicana*	*Piaya cayana mexicana*
48	*Piaya cayana thermophila*	
44	*Coccycua rutila panamensis*	*Piaya rutila panamensis*
21	*Coccyzus m. minor*	
19	*Coccyzus americanus julieni*	*Coccyzus a. americanus*
12	*Coccyzus a. americanus*	

Vol. VI		
784	*Glaucidium jardinii*	*Glaucidium j. jardinii*
765	*Ciccaba v. virgata*	
734	*Lophostrix stricklandi*	*Lophostrix cristata stricklandi*
688	*Otus a. asio*	
647	*Strix fulvescens*	
624	*Cryptoglaux tengmalmi richardsoni*	*Aegolius funereus richardsoni*
506	*Antrostomus carolinensis*	*Caprimulgus carolinensis*
485	*Hylomanes m. momotula*	
478	*Eumomota s. superciliosa*	
472	*Electron platyrhynchus minor*	*Electron platyrhynchum minor*
468	*Urospatha martii semirufa*	*Baryphthengus ruficapillus semirufus*
466	*Momotus castaneiceps*	*Momotus mexicanus castaneiceps*
463	*Momotus m. mexicanus*	
462	*Momotus subrufescens conexus*	*Momotus momota conexus*
457	*Momotus l. lessoni*	*Momotus momota lessonii*
449	*Todus todus*	
448	*Todus mexicanus*	
446	*Todus angustirostris*	
443	*Todus multicolor*	
440	*Chloroceryle aenea stictoptera*	
437	*Chloroceryle a. aenea*	
435	*Chloroceryle inda*	*Chloroceryle i. inda*
429	*Chloroceryle americana isthmica*	
424	*Chloroceryle amazona*	*Chloroceryle a. amazona*
415	*Streptoceryle a. alcyon*	*Megaceryle a. alcyon*
409	*Streptoceryle t. torquata*	*Megaceryle t. torquata*
401	*Nonnula frontalis*	*Nonnula f. frontalis*
398	*Monasa minor*	*Monasa morphoeus pallescens*
385	*Ecchaunornis radiatus fulvidus*	*Nystalus radiatus fulvidus*
382	*Hypnelus r. ruficollis*	
370	*Brachygalba salmoni*	
366	*Galbula melanogenia*	*Galbula ruficauda melanogenia*
363	*Jacamerops aurea*	*Jacamerops a. aurea*
357	*Aulacorhynchus c. caeruleogularis*	*Aulacorhynchus prasinus caeruleogularis*
355	*Aulacorhynchus p. prasinus*	
354	*Aulacorhynchus wagleri*	*Aulacorhynchus prasinus wagleri*

215

Vol. VI	Ridgway Nomenclature	Recent Nomenclature
350	*Selenidera spectabilis*	
347	*Pteroglossus sanguineus*	
345	*Pteroglossus frantzii*	
342	*Pteroglossus t. torquatus*	
337	*Ramphastos swainsonii*	
332	*Ramphastos p. piscivorus*	*Ramphastos s. sulfuratus*
317	*Eubucco bourcieri salvini*	*Eubucco bourcierii salvini*
309	*Nesoctites micromegas*	*Nesoctites m. micromegas*
304	*Picumnus olivaceus panamensis*	*Picumnus olivaceus flavotinctus*
286	*Sphyrapicus thyroideus*	*Sphyrapicus t. thyroideus*
282	*Sphyrapicus r. ruber*	*Sphyrapicus varius ruber*
274	*Sphyrapicus v. varius*	
261	*Dryobates a. arizonae*	*Dendrocopos a. arizonae*
228	*Dryobates p. pubescens*	*Dendrocopos p. pubescens*
218	*Dryobates villosus harrisi*	*Dendrocopos villosus harrisi*
201	*Dryobates v. villosus*	*Dendrocopos v. villosus*
184	*Xiphidiopicus p. percussus*	
181	*Cniparchus haematogaster splendens*	*Phloeoceastes haematogaster splendens*
174	*Scapaneus g. guatemalensis*	*Phloeoceastes g. guatemalensis*
155	*Phloeotomus p. pileatus*	*Dryocopus p. pileatus*
143	*Celeus l. loricatus*	
141	*Celeus castaneus*	
131	*Chloronerpes rubiginosus yucatanensis*	*Piculus rubiginosus yucatanensis*
129	*Chloronerpes aeruginosus*	*Piculus aeruginosus*
119	*Tripsurus p. pucherani*	*Melanerpes p. pucherani*
102	*Balanosphyra f. formicivora*	*Melanerpes f. formicivorus*
89	*Centurus c. chrysogenys*	*Melanerpes c. chrysogenys*
81	*Centurus aurifrons*	*Melanerpes a. aurifrons*
78	*Centurus p. polygrammus*	*Melanerpes aurifrons polygrammus*
71	*Centurus r. rubriventris*	*Melanerpes rubricapillus rubricomus*
42	*Melanerpes erythrocephalus*	*Melanerpes e. erythrocephalus*
37	*Colaptes mexicanoides*	*Colaptes cafer mexicanoides*
14	*Colaptes a. auratus*	

Vol. V

787	*Chrysotrogon caligatus*	*Trogon violaceus caligatus*
781	*Trogonurus curucui tenellus*	*Trogon rufus tenellus*
779	*Trogonurus a. aurantiiventris*	*Trogon a. aurantiiventris*
776	*Trogonurus puella*	*Trogon collaris puella*
774	*Trogonurus elegans*	*Trogon e. elegans*
768	*Trogonurus a. ambiguus*	*Trogon elegans ambiguus*
765	*Trogonurus mexicanus*	*Trogon m. mexicanus*
756	*Trogon m. melanocephalus*	*Trogon citreolus melanocephalus*
753	*Trogon bairdii*	*Trogon strigilatus bairdii*
749	*Curucujus clathratus*	*Trogon clathratus*
748	*Curucujus melanurus macrourus*	*Trogon melanurus macroura*
744	*Curucujus massena*	*Trogon m. massena*

Vol. V	Ridgway Nomenclature	Recent Nomenclature
741	*Leptuas neoxenus*	*Euptilotis neoxenus*
736	*Pharomachrus m. mocinno*	
694	*Tachornis p. phoenicobia*	
666	*Chrysolampis mosquitus*	
652	*Calothorax lucifer*	
645	*Nesophlox bryantae*	*Philodice bryantae*
629	*Archilochus colubris*	
625	*Calypte helenae*	
623	*Calypte costae*	
619	*Calypte anna*	
616	*Selasphorus floresii*	Hybrid: *Selasphorus sasin* × *Calypte anna*
612	*Selasphorus rufus*	
610	*Selasphorus alleni*	*Selasphorus s. sasin*
605	*Selasphorus ardens*	
592	*Atthis h. heloisa*	
489	*Lamprolaima rhami*	*Lamprolaima r. rhami*
420	*Amazilis bangsi*	*Amazilia bangsi* (hybrid: *A. r. rutila* × *A. t. tzacatl*)
419	*Amazilis rutila corallirostris*	*Amazilia rutila corallirostris*
416	*Amazilis r. rutila*	*Amazilia r. rutila*
414	*Amazilis yucatanensis cerviniventris*	*Amazilia yucatanensis cerviniventris*
413	*Amazilis y. yucatanensis*	*Amazilia y. yucatanensis*
350	*Anthoscenus c. constantii*	*Heliomaster c. constantii*
346	*Anthoscenus l. longirostris*	*Heliomaster l. longirostris*
324	*Phoethornis anthophilus hyalinus*	*Phaethornis anthophilus hyalinus*
319	*Phoethornis l. longirostris*	*Phaethornis superciliosus longirostris*
293	*Dendrocincla h. homochroa*	
291	*Dendrocincla lafresnayei ridgwayi*	*Dendrocincla fuliginosa ridgwayi*
288	*Dendrocincla a. anabatina*	
285	*Deconychura typica*	*Deconychura longicauda typica*
280	*Sittasomus s. sylvioides*	*Sittasomus griseicapillus sylvioides*
275	*Glyphorynchus cuneatus pectoralis*	*Glyphorynchus spirurus pectoralis*
272	*Campylorhamphus borealis*	*Campylorhamphus pusillus borealis*
271	*Campylorhamphus venezuelensis*	*Campylorhamphus trochilirostris venezuelensis*
264	*Picolaptes l. lineaticeps*	*Lepidocolaptes souleyetii lineaticeps*
261	*Picolaptes a. affinis*	*Lepidocolaptes a. affinis*
259	*Picolaptes leucogaster*	*Lepidocolaptes l. leucogaster*
254	*Xiphorhynchus erythropygius*	*Xiphorhynchus e. erythropygius*
250	*Xiphorhynchus n. nanus*	*Xiphorhynchus guttatus nanus*
249	*Xiphorhynchus striatigularis*	
244	*Xiphorhynchus f. flavigaster*	
237	*Xiphocolaptes e. emigrans*	*Xiphocolaptes promeropirhynchus emigrans*
233	*Dendrocolaptes puncticollis*	*Dendrocolaptes picumnus puncticollis*

217

Vol. V	Ridgway Nomenclature	Recent Nomenclature
233	*Dendrocolaptes validus costaricensis*	*Dendrocolaptes picumnus costaricensis*
229	*Dendrocolaptes s. sancti-thomae*	*Dendrocolaptes certhia sanctithomae*
220	*Automolus p. pallidigularis*	*Automolus ochrolaemus pallidigularis*
217	*Automolus c. cervinigularis*	*Automolus ochrolaemus cervinigularis*
216	*Automolus guerrerensis*	*Automolus rubiginosus guerrerensis*
214	*Automolus v. veraepacis*	*Automolus rubiginosus veraepacis*
209	*Xenicopsis subalaris lineatus*	*Syndactyla subalaris lineata*
207	*Xenicopsis variegaticeps*	*Anabacerthia striaticollis variegaticeps*
204	*Philydor fuscipennis*	*Philydor erythrocercus fuscipennis*
203	*Philydor panerythrus*	*Philydor rufus panerythrus*
194	*Synallaxis albescens latitabunda*	
189	*Synallaxis erythrothorax*	*Synallaxis e. erythrothorax*
185	*Acrorchilus erythrops rufigenis*	*Cranioleuca erythrops rufigenis*
181	*Premnoplex brunnescens brunneicauda*	
178	*Margarornis rubiginosa*	*Margarornis r. rubiginosa*
175	*Xenops rutilus heterurus*	
172	*Xenops genibarbis mexicanus*	*Xenops minutus mexicanus*
166	*Sclerurus canigularis*	*Sclerurus albigularis canigularis*
145	*Grallaricula costaricensis*	*Grallaricula flavirostris costaricensis*
135	*Phaenostictus m. mcleannani*	
132	*Anoplops bicolor*	*Gymnopithys leucaspis bicolor*
132	*Anoplops olivascens*	*Gymnopithys leucaspis olivascens*
128	*Hylophylax naevioides*	*Hylophylax n. naevioides*
121	*Formicarius moniliger intermedius*	*Formicarius analis intermedius*
121	*Formicarius moniliger pallidus*	*Formicarius analis pallidus*
120	*Formicarius m. moniliger*	*Formicarius analis moniliger*
119	*Formicarius analis nigricapillus*	*Formicarius n. nigricapillus*
114	*Myrmeciza zeledoni*	*Mermeciza immaculata zeledoni*
113	*Myrmeciza exsul occidentalis*	
111	*Myrmeciza e. exsul*	
110	*Myrmeciza cassini*	*Myrmeciza exsul cassini*
107	*Myrmeciza boucardi panamensis*	*Myrmeciza longipes panamensis*
102	*Gymnocichla chiroleuca*	*Gymnocichla nudiceps chiroleuca*
93	*Cercomacra t. tyrannina*	
85	*Ramphocaenus r. rufiventris*	*Ramphocaenus melanurus rufiventris*
62	*Myrmotherula surinamensis*	*Myrmotherula s. surinamensis*
59	*Dysithamnus striaticeps*	
58	*Dysithamnus puncticeps*	*Dysithamnus p. puncticeps*
56	*Dysithamnus mentalis septentrionalis*	
50	*Erionotus punctatus atrinucha*	*Thamnophilus punctatus atrinucha*
45	*Thamnophilus multistriatus*	*Thamnophilus m. multistriatus*

Vol. V	Ridgway Nomenclature	Recent Nomenclature
40	*Thamnophilus doliatus mexicanus*	*Thamnophilus doliatus intermedius*
37	*Thamnophilus radiatus nigricristatus*	*Thamnophilus doliatus nigricristatus*
34	*Hypolophus canadensis pulchellus*	*Sakesphorus canadensis pulchellus*
27	*Taraba t. transandeana*	*Taraba major transandeanus*
22	*Thamnistes a. anabatinus*	
20	*Cymbilaimus lineatus fasciatus*	

Vol. IV		
883	*Procnias tricarunculata*	
873	*Tityra semifasciata griseiceps*	
870	*Tityra semifasciata costaricensis*	
863	*Erator albitorques*	*Erator (Tityra) inquisitor fraserii*
849	*Platypsaris niger*	
840	*Pachyrhamphus cinnamomeus*	*Pachyrhamphus c. cinnamomeus*
838	*Pachyrhamphus cinereus*	*Pachyrhamphus r. rufus*
836	*Pachyrhamphus a. albo-griseus*	*Pachyrhamphus a. albogriseus*
829	*Pachyrhamphus polychropterus cinereiventris*	*Pachyrhamphus polychopterus cinereiventris*
823	*Lathria u. unirufa*	*Lipaugus u. unirufus*
820	*Lipaugus h. holerythrus*	*Rhytipterna h. holerythra*
811	*Attila citreopygus gaumeri*	*Attila spadiceus gaumeri*
804	*Attila tephrocephalus*	*Attila spadiceus citreopygus*
794	*Tyrannus elatus reguloides*	*Tyrannulus e. elatus*
792	*Microtriccus brunneicapillus*	*Ornithion semiflavum brunneicapillum*
791	*Microtriccus semiflavus*	*Ornithion s. semiflavus*
789	*Carpodectes antoniae*	
788	*Carpodectes nitidus*	
768	*Piprites griseiceps*	
765	*Laniocera rufescens*	*Laniocera r. rufescens*
762	*Scotothorus amazonus stenorhynchus*	*Schiffornis turdinus stenorhynchus*
759	*Scotothorus verae-pacis dumicola*	*Schiffornis turdinus veraepacis*
755	*Corapipo leucorrhoa altera*	
748	*Pipra e. erythrocephala*	
747	*Pipra mentalis ignifera*	
746	*Pipra m. mentalis*	
739	*Chiroprion linearis*	*Chiroxiphia pareola linearis*
737	*Chiroprion lanceolata*	*Chiroxiphia pareola lanceolata*
734	*Manacus aurantiacus*	*Manacus vitellinus aurantiacus*
732	*Manacus vitellinus*	*Manacus v. vitellinus*
731	*Manacus candei*	
712	*Tyrannus crassirostris*	*Tyrannus c. crassirostris*
706	*Tyrannus d. dominicensis*	
705	*Tyrannus melancholicus couchii*	
700	*Tyrannus melancholicus satrapa*	*Tyrannus melancholicus despotes*
697	*Tyrannus verticalis*	
694	*Tyrannus vociferans*	*Tyrannus v. vociferans*
675	*Pitangus lictor*	*Pitangus l. lictor*

Vol. IV	Ridgway Nomenclature	Recent Nomenclature
672	*Pitangus sulphuratus derbianus*	
669	*Coryphotriccus albovittatus*	*Conopias parva albovittata*
665	*Megarynchus pitangua mexicanus*	
662	*Myiodynastes hemichrysus*	*Myiodynastes chrysocephalus hemichrysus*
657	*Myiodynastes luteiventris*	*Myiodynastes l. luteiventris*
652	*Myiarchus barbirostris*	
651	*Myiarchus nigriceps*	*Myiarchus tuberculifer nigriceps*
647	*Myiarchus lawrencii querulus*	*Myiachus tuberculifer querulus*
642	*Myiarchus l. lawrenceii*	*Myiarchus tuberculifer lawrencei*
638	*Myiarchus antillarum*	
634	*Myiarchus dominicensis*	
633	*Myiarchus stolidus*	*Myiarchus s. stolidus*
632	*Myiarchus yucatanensis*	
629	*Myiarchus n. nuttingi*	
625	*Myiarchus c. cinerascens*	
621	*Myiarchus m. mexicanus*	*Myiarchus tyrannulus nelsoni*
617	*Myiarchus o. oberi*	
613	*Myiarchus crinitus*	*Myiarchus c. crinitus*
606	*Eribates magnirostris*	*Myiarchus magnirostris*
594	*Sayornis phoebe*	
591	*Empidonax fulvifrons rubicundus*	
588	*Empidonax f. fulvifrons*	
586	*Empidonax atriceps*	
584	*Empidonax albigularis*	*Empidonax a. albigularis*
583	*Empidonax flavescens*	*Empidonax f. flavescens*
582	*Empidonax salvini*	*Empidonax flavescens salvini*
581	*Empidonax difficilis bairdi*	*Empidonax difficilis occidentalis*
576	*Empidonax d. difficilis*	
575	*Empidonax trepidus*	*Empidonax affinis trepidus*
572	*Empidonax pulverius*	*Empidonax affinis pulverius*
570	*Empidonax griseus*	
567	*Empidonax wrightii*	*Empidonax oberholseri*
561	*Empidonax minimus*	
558	*Empidonax traillii alnorum*	*Empidonax alnorum*
555	*Empidonax t. traillii*	
552	*Empidonax virescens*	
549	*Empidonax flaviventris*	
543	*Myiophobos fasciatus furfurosus*	
537	*Blacicus blancoi*	*Contopus latirostris blancoi*
536	*Blacicus pallidus*	*Contopus caribaeus pallidus*
535	*Blacicus hispaniolensis*	*Contopus caribaeus hispaniolensis*
521	*Myiochanes r. richardsonii*	*Contopus virens richardsonii*
518	*Myiochanes virens*	*Contopus v. virens*
515	*Myiochanes pertinax pallidiventris*	*Contopus fumigatus pallidiventris*
513	*Myiochanes p. pertinax*	*Contopus fumigatus pertinax*
505	*Nuttallornis borealis*	
501	*Mitrephanes aurantiiventris*	*Mitrephanes phaeocercus aurantiiventris*
498	*Mitrephanes p. phaeocercus*	

Vol. IV	Ridgway Nomenclature	Recent Nomenclature
495	*Terenotriccus erythrurus fulvigularis*	
493	*Aphanotriccus capitalis*	
490	*Myiobius xanthopygus sulphureipygius*	*Myiobius barbatus sulphureipygius*
488	*Myobius barbatus atricaudus*	*Myiobius a. atricaudus*
484	*Cnipodectes subbrunneus*	*Cnipodectes s. subbrunneus*
475	*Pyrocephala rubinus mexicanus*	
469	*Capsiempis flaveola*	*Capsiempis f. flaveola*
467	*Leptopogon flavovirens*	*Phylloscartes ventralis flavovirens*
465	*Leptopogon superciliaris*	*Leptopogon s. superciliaris*
463	*Leptopogon pileatus*	*Leptopogon amaurocephalus pileatus*
461	*Mionectes o. olivaceus*	
459	*Pipromorpha semischistacea*	*Pipromorpha oleaginea assimilis*
457	*Pipromorpha oleaginea parca*	
455	*Pipromorpha assimilis dyscola*	*Pipromorpha oleaginea dyscola*
454	*Pipromorpha a. assimilis*	
450	*Myiozetetes granadensis*	*Myiozetetes g. granadensis*
447	*Myiozetetes t. texensis*	*Myiozetetes similis texensis*
444	*Myiozetetes c. cayanensis*	
434	*Elaenia f. frantzii*	
429	*Elaenia martinica subpagana*	*Elaenia flavogaster subpagana*
426	*Elaenia m. martinica*	
422	*Sublegatus glaber*	*Sublegatus arenarum glaber*
417	*Camptostoma pusillum flaviventre*	*Camptostoma obsoletum flaviventre*
414	*Camptostoma imberbe*	*Camptostoma i. imberbe*
408	*Tyranniscus v. vilissimus*	
405	*Myiopagis cotta*	
404	*Myiopagis p. placens*	*Myiopagis viridicata placens*
394	*Rhynchocyclus cinereiceps*	*Tolmomyias sulphurescens cinereiceps*
392	*Rhynchocyclus marginatus*	*Tolmomyias assimilis flavotectus*
391	*Rhynchocyclus flavo-olivaceus*	*Tolmomyias sulphurescens flavo-olivaceus*
388	*Craspedoprion brevirostris*	*Rhynchocyclus b. brevirostris*
387	*Craspedoprion aequinoctialis*	*Rhynchocyclus olivaceus aequinoctialis*
384	*Platytriccus albogularis*	*Platyrhinchus mystaceus albogularis*
382	*Platytriccus cancrominus*	*Platyrhinchus mystaceus cancrominus*
379	*Placostomus superciliaris*	*Platyrhinchus coronatus superciliaris*
377	*Perissotriccus atricapillus*	*Myiornis ecaudatus atricapillus*
374	*Atalotriccus p. pilaris*	
371	*Lophotriccus squamaecristatus minor*	*Lophotriccus pileatus squamaecristatus*
366	*Todirostrum nigriceps*	*Todirostrum chrysocrotaphum nigriceps*

221

Vol. IV	Ridgway Nomenclature	Recent Nomenclature
367	*Todirostrum schistaceiceps*	*Todirostrum sylvia schistaceiceps*
364	*Todirostrum cinerum finitimum*	
360	*Oncostoma olivaceum*	*Oncostoma cinereigulare olivaceum*
358	*Oncostoma cinereigulare*	*Oncostoma c. cinereigulare*
354	*Onychorhynchus m. mexicanus*	
334	*Oxyruncus cristatus frater*	
325	*Otocoris alpestris adusta*	
320	*Otocoris alpestris actia*	
316	*Otocoris alpestris strigata*	
311	*Otocoris alpestris praticola*	
303	*Octocoris a. alpestris*	*Eremophila a. alpestris*
287	*Estrilda melpoda*	*Estrilda m. melpoda*
285	*Spermestes cucullata*	*Lonchura c. cucullata*
269	*Ramphocinclus brachyurus*	
259	*Oroscoptes montanus*	
241	*Mimus g. gundlachii*	
234	*Mimus g. gilvus*	
225	*Mimus p. polyglottos*	
218	*Galeoscoptes carolinensis*	*Dumetella carolinensis*
216	*Mimodes graysoni*	
203	*Toxostoma redivivum*	*Toxostoma r. redivivum*
201	*Toxostoma curvirostre occidentale*	
199	*Toxostoma c. curvirostre*	
197	*Toxostoma bendirei*	*Toxostoma b. bendirei*
195	*Toxostoma c. cinereum*	
187	*Toxostoma rufum*	*Tostoma r. rufum*
179	*Myadestes sibilans*	
175	*Myadestes g. genibarbis*	
172	*Myadestes elizabeth*	*Myadestes e. elizabeth*
166	*Myadestes o. obscurus*	
163	*Myadestes townsendi*	*Myadestes t. townsendi*
156	*Sialia arctica*	*Sialia currucoides*
152	*Sialia mexicana bairdi*	
148	*Sialia m. mexicana*	
146	*Sialia sialis fulva*	
142	*Sialia s. sialis*	
136	*Ridgwayia pinicola*	*Zoothera pinicola*
117	*Planesticus g. grayi*	*Turdus g. grayi*
115	*Planesticus nigrirostris*	*Turdus fumigatus bondi*
113	*Planesticus gymnophthalmus*	*Turdus albicollis phaeopygus*
97	*Planesticus m. migratorius*	*Turdus m. migratorius*
74	*Cichlherminia h. herminieri*	*Cichlherminia l. lherminieri*
67	*Hylocichla fuscescens salicicola*	*Catharus fuscescens salicicola*
64	*Hylocichla f. fuscescens*	*Catharus f. fuscescens*
59	*Hylocichla a. aliciae*	*Catharus m. minimus*
55	*Hylocichla ustulata swainsonii*	*Catharus ustulatus swainsoni*
52	*Hylocichla u. ustulata*	*Catharus u. ustulatus*
39	*Hylocichla g. guttata*	*Catharus g. guttatus*
37	*Hylocichla mustelina*	
32	*Catharus g. gracilirostris*	

Vol. IV	Ridgway Nomenclature	Recent Nomenclature
29	*Catharus m. melpomene*	*Catharus aurantiirostris melpomene*
28	*Catharus f. frantzii*	*Catharus occidentalis frantzii*
26	*Catharus o. occidentalis*	
24	*Catharus mexicanus fumosus*	
22	*Catharus m. mexicanus*	
15	*Cyanosylvia suecica*	*Erithacus s. suecicus*

Vol. III		
705	*Regulus c. calendula*	
686	*Chamaea f. fasciata*	
628	*Thryophilus thoracicus*	*Thryophilus t. thoracicus*
624	*Thryophilus c. castaneus*	*Thryothorus nigricapillus castaneus*
602	*Olbiorchilus alascensis*	*Troglodytes troglodytes alascensis*
597	*Olbiorchilus h. hiemalis*	*Troglodytes troglodytes hiemalis*
593	*Troglodytes martinicensis*	*Troglodytes aëdon martinicensis*
590	*Troglodytes ochraceus*	*Troglodytes solstitialis ochraceus*
579	*Troglodytes a. aëdon*	
541	*Thryothorus l. ludovicianus*	
534	*Pheugopedius m. maculipectus*	*Thryothorus m. maculipectus*
533	*Pheugopedius hyperythrus*	*Thryothorus rutilus hyperythrus*
532	*Pheugopedius fasciato-ventris melanogaster*	*Thryothorus fasciatoventris melanogaster*
531	*Pheugopedius fasciato-ventris albigularis*	*Thryothorus fasciatoventris albigularis*
514	*Heleodytes nelsoni*	*Campylorhynchus megalopterus nelsoni*
507	*Heleodytes rufinucha*	*Campylorhynchus r. rufinucha*
505	*Heleodytes c. capistratus*	*Campylorhynchus rufinucha capistratus*
504	*Heleodytes chiapensis*	*Campylorhynchus griseus chiapensis*
503	*Heleodytes albobrunneus*	*Campylorhynchus turdinus albobrunneus*
498	*Telmatodytes palustris thryophilus*	*Cistothorus palustris thryophilus*
456	*Sitta p. pygmaea*	
453	*Sitta pusilla*	
450	*Sitta canadensis*	
441	*Sitta c. carolinensis*	
432	*Psaltriparus m. minimus*	
431	*Psaltriparus plumbeus*	*Psaltriparus minimus plumbeus*
429	*Psaltriparus melanotis lloydi*	*Psaltriparus minimus lloydi*
426	*Psaltriparus m. melanotis*	*Psaltriparus m. minimus*
416	*Penthestes r. rufescens*	*Parus r. rufescens*
411	*Penthestes cinctus alascens*	*Parus cinctus lathami*
408	*Penthestes gambeli*	*Parus g. gambeli*
407	*Penthestes sclateri*	*Parus s. sclateri*
404	*Penthestes c. carolinensis*	*Parus c. carolinensis*
397	*Penthestes a. atricapillus*	*Parus a. atricapillus*
392	*Baeolophus w. wollweberi*	*Parus w. wollweberi*

Vol. III	Ridgway Nomenclature	Recent Nomenclature
390	*Baeolophus inornatus griseus*	*Parus inornatus ridgwayi*
361	*Cyanocitta stelleri azteca*	
347	*Cyanocitta c. cristata*	
344	*Aphelocoma u. unicolor*	
336	*Aphelocoma sumichrasti*	*Aphelocoma coerulescens sumichrasti*
327	*Aphelocoma c. californica*	*Aphelocoma coerulescens californica*
326	*Aphelocoma cyanea*	*Aphelocoma c. coerulescens*
315	*Cissilopha yucatanica*	*Cissilopha sanblasiana yucatanica*
313	*Cissilopha s. sanblasiana*	
306	*Xanthoura l. luxuosa*	*Cyanocorax yncas luxuosus*
281	*Nucifraga columbiana*	
237	*Lanius borealis*	*Lanius excubitor borealis*
230	*Cyclarhis flavipectus subflavescens*	*Cyclarhis gujanensis subflavescens*
228	*Cyclarhis f. flaviventris*	*Cyclarhis gujanensis flaviventris*
225	*Vireolanius melitophrys*	
223	*Vireolanius p. pulchellus*	
221	*Pachysylvia viridiflava*	*Hylophilus flavipes viridiflavus*
220	*Pachysylvia a. aurantiifrons*	*Hylophilus a. aurantiifrons*
218	*Pachysylvia o. ochraceiceps*	*Hylophilus o. ochraceiceps*
216	*Pachysylvia decurtata*	*Hylophilus d. decurtatus*
210	*Vireo latimeri*	
204	*Vireo b. bellii*	
201	*Vireo h. hypochryseus*	
199	*Vireo carmioli*	
197	*Vireo huttoni stephensi*	
195	*Vireo h. huttoni*	
194	*Vireo pallens*	*Vireo p. pallens*
188	*Vireo gundlachi*	
181	*Vireo atricapillus*	
167	*Lanivireo s. solitarius*	*Vireo s. solitarius*
166	*Lanivireo propinquus*	Hybrid: *Vireo flavifrons* × *V. solitarius*
163	*Lanivireo flavifrons*	*Vireo flavifrons*
153	*Vireosylva g. gilva*	*Vireo g. gilvus*
151	*Vireosylva philadelphica*	*Vireo philadelphicus*
147	*Vireosylva olivacea*	*Vireo olivaceus*
144	*Vireosylva f. flavoviridis*	*Vireo olivaceus flavoviridis*
136	*Vireosylva caymanensis*	*Vireo magister caymanensis*
117	*Ptilogonys cinereus molybdophanes*	
116	*Ptilogonys c. cinereus*	
109	*Ampelis cedrorum*	*Bombycilla cedrorum*
105	*Ampelis garrulus*	*Bombycilla g. garrulus*
99	*Callichelidon cyaneoviridis*	
80	*Hirundo erythrogastra*	*Hirundo rustica erythrogaster*
79	*Hirundo rustica*	*Hirundo r. rustica*
66	*Notiochelidon pileata*	
58	*Stelgidopteryx serripennis*	*Stelgidopteryx ruficollis serripennis*

Vol. III	Ridgway Nomenclature	Recent Nomenclature
9	*Budytes flavus alascensis*	*Motacilla flava tschutschensis*

Vol. II		
772	*Rhodinocichla schistacea*	*Rhodinocichla rosea schistacea*
771	*Rhodinocichla rosea eximia*	
760	*Ergaticus versicolor*	
759	*Ergaticus ruber*	*Ergaticus r. ruber*
753	*Basileuterus c. culicivorus*	
751	*Basileuterus melanogenys*	*Basileuterus m. melanogenys*
745	*Basileuterus r. rufifrons*	
743	*Basileuterus b. belli*	
731	*Myioborus m. miniatus*	
728	*Setophaga p. picta*	
720	*Cardellina rubrifrons*	
710	*Wilsonia p. pusilla*	
712	*Wilsonia pusilla pileolata*	
705	*Wilsonia mitrata*	*Wilsonia citrina*
701	*Granatellus s. sallaei*	
699	*Granatellus venustus*	*Granatellus v. venustus*
692	*Icteria v. virens*	
687	*Chamaethlypis p. poliocephala*	*Geothlypis p. poliocephala*
674	*Geothlypis rostrata*	*Geothlypis r. rostrata*
673	*Geothlypis trichas melanops*	
664	*Geothlypis trichas brachidactyla*	
661	*Geothlypis t. trichas*	
642	*Seiurus n. noveboracensis*	
639	*Seiurus motacilla*	
631	*Oporornis tolmiei*	*Oporornis t. tolmiei*
628	*Oporornis philadelphia*	
626	*Oporornis agilis*	
623	*Oporornis formosa*	*Oporornis formosus*
612	*Dendroica p. palmarum*	
611	*Dendroica vitellina*	*Dendroica v. vitellina*
607	*Dendroica discolor*	
604	*Dendroica kirtlandi.*	
599	*Dendroica v. vigorsii*	*Dendroica p. pinus*
595	*Dendroica striata*	
592	*Dendroica castanea*	
589	*Dendroica pensylvanica*	
584	*Dendroica g. graciae*	
579	*Dendroica d. dominica*	
574	*Dendroica blackburniae*	*Dendroica fusca*
570	*Dendroica rara*	*Dendroica cerulea*
565	*Dendroica chrysoparia*	
562	*Dendroica virens*	
559	*Dendroica townsendi*	
546	*Dendroica coronata*	*Dendroica c. coronata*
537	*Dendroica tigrina*	
533	*Dendroica maculosa*	*Dendroica magnolia*
529	*Dendroica b. bryanti*	*Dendroica petechia bryanti*

225

Vol. II	Ridgway Nomenclature	Recent Nomenclature
527	*Dendroica erithachorides*	*Dendroica petechia erithachorides*
526	*Dendroica rufigula*	*Dendroica petechia ruficapilla*
523	*Dendroica r. ruficapilla*	*Dendroica petechia ruficapilla*
515	*Dendroica p. petechia*	
508	*Dendroica a. aestiva*	*Dendroica petechia aestiva*
494	*Peucedramus olivaceus*	*Peucedramus t. taeniatus*
478	*Oreothlypis superciliosa*	*Vermivora s. superciliosa*
473	*Helminthophila crissalis*	*Vermivora crissalis*
471	*Helminthophila virginiae*	*Vermivora virginiae*
468	*Helminthophila r. rubricapilla*	*Vermivora r. ruficapilla*
466	*Helminthophila celata lutescens*	*Vermivora celata lutescens*
462	*Helminthophila c. celata*	*Vermivora c. celata*
460	*Helminthophila peregrina*	*Vermivora peregrina*
458	*Helminthophila bachmani*	*Vermivora bachmanii*
455	*Helminthophila pinus*	*Vermivora pinus*
452	*Helminthophila lawrencii*	Hybrid: *Vermivora pinus* × *V. chrysoptera*
442	*Protonotaria citrea*	
439	*Helmitheros vermivorus*	
436	*Helinaia swainsonii*	*Limnothlypis swainsonii*
421	*Coereba uropygialis*	*Coereba flaveola uropygialis*
420	*Coereba barbadensis*	*Coereba flaveola barbadensis*
415	*Coereba saccharina*	*Coereba flaveola atrata*
409	*Coereba mexicana*	*Coereba flaveola mexicana*
408	*Coereba cerinoclunis*	*Coereba flaveola cerinoclunis*
401	*Coereba bahamensis*	*Coereba flaveola bahamensis*
397	*Dacnis venusta*	*Dacnis v. venusta*
386	*Cyanerpes cyaneus*	*Cyanerpes c. cyaneus*
383	*Chlorophanes spiza guatemalensis*	
381	*Diglossa plumbea*	*Diglossa baritula plumbea*
380	*Diglossa baritula*	*Diglossa b. baritula*
370	*Dolichonyx oryzivorus*	
357	*Sturnella m. magna*	
330	*Agelaius p. phoeniceus*	
326	*Agelaius g. gubernator*	*Agelaius phoeniceus gubernator*
315	*Icterus bullockii*	*Icterus galbula bullockii*
311	*Icterus galbula*	*Icterus g. galbula*
309	*Icterus parisorum*	
305	*Icterus m. mesomelas*	
303	*Icterus leucopteryx*	*Icterus l. leucopteryx*
301	*Icterus x. xanthornus*	*Icterus n. nigrogularis*
299	*Icterus auratus*	
291	*Icterus cucullatus igneus*	
288	*Icterus c. cucullatus*	
285	*Icterus g. gularis*	
283	*Icterus p. pectoralis*	
279	*Icterus bonana*	
275	*Icterus spurius*	*Icterus s. spurius*
271	*Icterus hypomelas*	*Icterus dominicensis*
243	*Megaquiscalus tenuirostris*	*Quiscalus palustris*

226

Vol. II	Ridgway Nomenclature	Recent Nomenclature
238	*Megaquiscalus major macrourus*	*Quiscalus m. mexicanus*
237	*Megaquiscalus m. major*	*Quiscalus m. major*
217	*Quiscalus quiscula oglaeus*	*Quiscalus q. quiscula*
207	*Molothrus a. ater*	
191	*Cassiculus melanicterus*	*Cacicus melanicterus*
180	*Gymnostinops montezuma*	*Psarocolius montezuma*
167	*Chlorospingus hypophaeus*	*Chlorospingus flavigularis hypophaeus*
166	*Chlorospingus punctulatus*	
165	*Chlorospingus pileatus*	*Chlorospingus p. pileatus*
163	*Chlorospingus albitempora*	*Chlorospingus ophthalmicus regionalis*
161	*Chlorospingus ophthalmicus*	*Chlorospingus o. ophthalmicus*
155	*Chlorothraupis carmioli*	*Chlorothraupis c. carmioli*
154	*Chlorothraupis olivaceus*	
152	*Phoenicothraupis fuscicauda*	*Habia f. fuscicauda*
151	*Phoenicothraupis salvini peninsularis*	*Habia fuscicauda insularis*
149	*Phoenicothraupis salvini littoralis*	*Habia fuscicauda salvini*
148	*Phoenicothraupis s. salvini*	*Habia fuscicauda salvini*
144	*Phoenicothraupis rubica rubicoides*	*Habia rubica rubicoides*
140	*Eucometis s. spodocephala*	*Eucometis penicillata spodocephala*
139	*Eucometis cristata*	*Eucometis penicillata cristata*
136	*Tachyphonus delatrii*	
134	*Tachyphonus axillaris*	*Tachyphonus luctuosus axillaris*
133	*Tachyphonus lactuosus*	*Tachyphonus l. lactuosus*
128	*Phaenicophilus poliocephalus*	*Phaenicophilus p. poliocephalus*
127	*Phaenicophilus palmarum*	
124	*Lanio leucothorax*	*Lanio l. leucothorax*
125	*Lanio melanopygius*	*Lanio leucothorax melanopygius*
116	*Ramphocelus d. dimidratus*	
115	*Ramphocelus luciani*	*Ramphocelus m. melanogaster*
111	*Ramphocelus costaricensis*	*Ramphocelus passerinii costaricensis*
109	*Ramphocelus passerinii*	*Ramphocelus p. passerinii*
106	*Hemithraupis chrysomelas*	*Chrysothlypis c. chrysomelus*
102	*Piranga erythrocephala*	*Piranga e. erythrocephala*
99	*Piranga l. leucoptera*	
98	*Piranga r. roseo-gularis*	*Piranga r. roseogularis*
95	*Piranga b. bidentata*	
88	*Piranga erythromelas*	*Piranga ludoviciana*
86	*Piranga t. testacea*	*Piranga flava testacea*
85	*Piranga hepatica*	*Piranga flava hepatica*
83	*Piranga rubra cooperi*	
79	*Piranga r. rubra*	
73	*Spindalis benedicti*	*Spindalis zena benedicti*
70	*Spindalis z. zena*	
69	*Spindalis pretrei*	*Spindalis zena pretrei*
67	*Spindalis multicolor*	*Spindalis zena dominicensis*
66	*Spindalis portoricensis*	*Spindalis zena portoricensis*

Vol. II	Ridgway Nomenclature	Recent Nomenclature
64	*Spindalis nigricephala*	*Spindalis zena nigricephala*
60	*Tanagra abbas*	*Thraupis abbas*
52	*Calospiza cucullata*	*Tangara c. cucullata*
46	*Calospiza lavinia*	*Tangara l. lavinia*
43	*Calospiza gyroloides*	*Tangara gyrola viridissima*
39	*Calospiza f. florida*	*Tangara f. florida*
31	*Pyrrhuphonia jamaica*	*Euphonia jamaica*
29	*Euphonia gouldi*	*Euphonia g. gouldi*
28	*Euphonia crassirostris*	*Euphonia laniirostris crassirostris*
25	*Euphonia hirundinacea*	
24	*Euphonia godmani*	*Euphonia affinis godmani*
23	*Euphonia minuta humilis*	
21	*Euphonia affinis*	*Euphonia a. affinis*
20	*Euphonia luteicapilla*	
19	*Euphonia gracilis*	*Euphonia imitans*
18	*Euphonia fulvicrissa*	
17	*Euphonia anneae*	*Euphonia a. anneae*
16	*Euphonia flavifrons*	*Euphonia musica flavifrons*
15	*Euphonia sclateri*	
14	*Euphonia musica*	
12	*Euphonia elegantissima*	

Vol. I		
669	*Saltator albicollis isthmicus*	
663	*Saltator m. magnoides*	*Saltator maximus magnoides*
666	*Caryothraustes poliogaster scapularis*	*Caryothraustes canadensis*
661	*Saltator a. atriceps*	
657	*Rhodothraupis celaeno*	
642	*Cardinalis cardinalis coccineus*	*Cardinalis c. coccineus*
635	*Cardinalis c. cardinalis*	
625	*Pyrrhuloxia s. sinuata*	
623	*Pheucticus aurantiacus*	*Pheucticus chrysopeplus aurantiacus*
622	*Pheucticus chrysopeplus*	*Pheucticus c. chrysopeplus*
614	*Zamelodia ludoviciana*	*Pheucticus ludovicianus*
607	*Guiraca c. caerulea*	*Passerina c. caerulea*
601	*Cyanocompsa p. parellina*	*Passerina p. parellina*
599	*Cyanocompsa cyanoides*	*Passerina c. cyanoides*
591	*Cyanospiza versicolor*	*Passerina v. versicolor*
590	*Cyanospiza rositae*	*Passerina rositae*
589	*Cyanospiza leclancheri*	*Passerina l. leclancherii*
586	*Cyanospiza ciris*	*Passerina c. ciris*
584	*Cyanospiza amoena*	*Passerina amoena*
582	*Cyanospiza cyanea*	*Passerina cyanea*
577	*Sporophila torqueola*	*Sporophila t. torqueola*
575	*Sporophila morelleti*	*Sporophila torqueola morelleti*
554	*Pyrrhulagra n. noctis*	*Loxigilla n. noctis*
546	*Loxipasser anoxanthus*	
538	*Euetheia bicolor omissa*	*Tiaris bicolor omissa*

228

Vol. I	Ridgway Nomenclature	Recent Nomenclature
537	*Euetheia b. bicolor*	*Tiaris b. bicolor*
536	*Euetheia canora*	*Tiaris canora*
526	*Volatinia jacarini splendens*	
523	*Sicalis chrysops*	*Sicalis luteola chrysops*
477	*Camarhynchus psittaculus*	*Camarhynchus p. psittacula*
472	*Pezopetes capitalis*	
468	*Buarremon assimilis*	*Atlapetes coronatus assimilis*
467	*Buarremon virenticeps*	*Atlapetes torquatus virenticeps*
465	*Buarremon brunneinuchus*	*Atlapetes b. brunneinucha*
463	*Atlapetes albinucha*	*Atlapetes a. albinucha*
461	*Atlapetes gutturalis*	*Atlapetes albinucha gutturalis*
458	*Lysurus crassirostris*	*Lysurus castineiceps crassirostris*
453	*Arremonops conirostris richmondi*	*Arremonops chloronotus richmondi*
452	*Arremonops chloronotus*	*Arremonops c. chloronotus*
449	*Arremonops s. superciliosus*	*Arremonops rufivirgatus superciliosus*
447	*Arremonops r. rufivirgatus*	
423	*Pipilo e. erythrophthalmus*	
406	*Pipilo t. torquatus*	Hybrid: *Pipilo ocai* × *P. erythrophthalmus*
395	*Passerella iliaca schistacea*	
311	*Spizella s. socialis*	*Spizella p. passerina*
303	*Junco alticola*	*Junco phaenotus alticola*
291	*Junco mearnsi*	*Junco hyemalis mearnsi*
251	*Aimophila ruficeps eremoeca*	
246	*Aimophila r. ruficeps*	
223	*Ammodramus caudacutus subvirgatus*	*Ammospiza caudacuta subvirgata*
214	*Ammodramus m. maritimus*	*Ammospiza m. maritima*
143	*Passer domesticus*	
131	*Carpodacus m. mexicanus*	
128	*Carpodacus p. purpureus*	
126	*Carpodacus cassinii*	
118	*Astragalinus psaltria croceus*	*Spinus psaltria colombianus*
114	*Astragalinus p. psaltria*	*Spinus p. psaltria*
109	*Astragalinus t. tristis*	*Spinus t. tristis*
107	*Loximitris dominicensis*	
104	*Spinus xanthogaster*	
102	*Spinus n. notatus*	
100	*Spinus atriceps*	
80	*Acanthis h. hornemannii*	
68	*Leucosticte t. tephrocotis*	
60	*Pinicola enucleater canadensis*	*Pinicola enucleater leucura*
53	*Loxia leucoptera*	*Loxia l. leucoptera*
44	*Hesperiphona abeillii*	*Hesperiphona abeillei*
39	*Hesperiphona v. vespertina*	

1673